JUDAISM AND ECOLOGY

WORLD RELIGIONS AND ECOLOGY

This series looks at how each of five world religions has treated ecology in the past, what the teaching of each has to say on the subject, and how that is applied today. Contributors from a variety of backgrounds in each religion put forward material for thought and discussion through poetry, stories, and pictures as well as ideas and theories.

The series is sponsored by the World Wide Fund for Nature, who believe that a true understanding of our relation to the natural world is the best step towards saving our planet.

Titles in the series are:

BUDDHISM AND ECOLOGY
*Martine Batchelor
and Kerry Brown*

CHRISTIANITY AND ECOLOGY
*Elizabeth Breuilly
and Martin Palmer*

HINDUISM AND ECOLOGY: Seeds of Truth
Ranchor Prime

ISLAM AND ECOLOGY
*Fazlun M. Khalid
with Joanne O'Brien*

JUDAISM AND ECOLOGY
Aubrey Rose

JUDAISM
AND
ECOLOGY

Edited by

Aubrey Rose

CASSELL

Cassell Publishers Limited
Villiers House, 41/47 Strand, London WC2N 5JE, England
387 Park Avenue South, New York, NY 10016–8810, USA

© World Wide Fund for Nature 1992

First published 1992

British Library Cataloguing-in-Publication Data
A catalogue record for this book is available from the British
Library.

Library of Congress Cataloging-in-Publication Data
Available from the Library of Congress.

ISBN 0–304–32378–0

Cover photo: ZEFA.
Panda symbol © 1986 World Wide Fund for Nature

Typeset by Fakenham Photosetting Limited, Fakenham, Norfolk
Printed and bound in Great Britain by Mackays of Chatham plc,
Chatham, Kent

CONTENTS

Foreword *Chief Rabbi Dr Jonathan Sacks*   vii

Introduction   ix

Glossary   xiii

Acknowledgements   xviii

BLESSINGS   2

1 A PERSONAL VIEW   4
  Aubrey Rose

SECTION A · RESOURCES IN JEWISH
TEACHING AND TRADITION

2 INTRODUCTION TO THE JEWISH FAITH   9
  Aubrey Rose

3 JUDAISM AND THE ENVIRONMENT   19
  Norman Solomon

4 TRADITIONAL JEWISH ATTITUDES TOWARDS   54
  PLANT AND ANIMAL CONSERVATION
  Yosef Orr and Yossi Spanier

5 ANIMAL WELFARE   61

6 TU BI SHEVAT: A HAPPY NEW YEAR TO ALL
  TREES! *Philip L. Pick*   67

7   SHEMITTA: A SABBATICAL FOR THE LAND   70
    'The Land shall rest and the people shall grow'
    Shlomo Riskin

8   THE SOURCES OF VEGETARIAN   74
    INSPIRATION *Philip L. Pick*

    SECTION B · ISRAEL

9   THE ENVIRONMENT: ISRAEL'S REMARKABLE   82
    STORY *Aubrey Rose*

10 THE REGIONAL AND GLOBAL SIGNIFICANCE   91
    OF ENVIRONMENTAL PROTECTION,
    NATURE CONSERVATION AND ECOLOGICAL
    RESEARCH IN ISRAEL *Uriel N. Safriel*

11 NATURE RESERVES IN ISRAEL *Aubrey Rose*   100

12 JERUSALEM'S BOTANICAL GARDEN   105
    Watching the dream come true

13 NOAH'S SANCTUARIES *Liat Collins*   109

    SECTION C · ACTION

14 THE USA: A JEWISH ECOLOGY GROUP   114

    BROWNIE PRAYERS ON THE ENVIRONMENT   117

15 ACTION ON THE ENVIRONMENT: 119
    A PRACTICAL GUIDE *Vicky Joseph*

16 A SYNAGOGUE ENVIRONMENTAL
    GROUP *Sheila Chiat*   128

    JUDAISM AND ECOLOGY: AN OVERVIEW OF
    SOURCE MATERIAL *Compiled by Sammy Jackman* 131

    BIBLIOGRAPHY   141

FOREWORD

I write this brief foreword with great pleasure. The subject of ecology is of paramount importance. Our Jewish contribution in this field, from the first words of the Bible to the present day, has been full of insight, practical advice, and deep concern for nature, human beings and all forms of life.

This is reflected in the many articles assembled and introduced by the editor, containing teaching and principles enunciated long ago, developed in talmudic times and by succeeding generations of rabbis, yet applicable to the changed and ever-changing environment in which we find ourselves today.

The publication of this book, indeed of the series of which it forms part, is a splendid and timely initiative which I am sure will be read with enormous interest by both the Jewish and non-Jewish public.

I commend both publishers and editor on their achievement.

Jonathan Sacks

Chief Rabbi Dr Jonathan Sacks

TO DAVID

INTRODUCTION

I would like to guide you through this book and its varied contents, surveying the environment from the viewpoint of Judaism. This could of course have been an impersonal religious statement on a crucial subject, but some explanation, I felt, was needed, however brief, of the writers, of special terms, expressions and doctrines, of historical development.

I also felt the book should not be heavy with theory—an unJudaic approach—and hence the reader will find interspersed an odd cartoon, possibly a joke or two (thank heavens Jews have had an ability to laugh at themselves, they needed to). A glossary at the beginning explains the meaning of some phrases transliterated from the original Hebrew, along with reference to key personalities, while at the end are sources for further study. A wealth of material presented itself, much of which I have had regretfully to exclude, but which will certainly find its way into a more extended and detailed study. Of course each contributor expresses his or her personal views.

But why me? We have enough technical experts, rabbinical and scientific, far more learned than I am, who could have taken my place. The simple answer is that I was asked to edit the book, and I did so gladly. Perhaps I was chosen as I have spoken and written widely on the subject, representing the views of British Jews before Parliamentary Committees, in inter-faith declarations, as at Canterbury in 1988. Perhaps because with the approval of the Board of Deputies of British Jews (founded 1760 and still the representative voice of British Jewry) I set up a committee of Jewish experts to advise the community on environmental matters, emphasizing Jewish religious teachings. I am immensely grateful to those experts and to my wife Sheila, a tower of strength, in the production of this book, as I am also grateful to the publishers Cassell and especially Ruth McCurry (my publishing mentor and co-worker) for allowing me to be your guide. A particular word of thanks is due to World Wide Fund for Nature for their unfailing support.

This is not a book that bursts on to the public in splendid isolation. It is part of a series, a commendable series, indicating teachings of the great faiths on the natural world around us. A word therefore on the relationship between Judaism and these faiths. Apart from Hinduism, Judaism is the oldest surviving religion, originating about 4,000 years ago, rising to its heights in biblical times, but constantly developing ever since, evolving its practice against a background of certain fundamental immutable statements.

It is a practical faith. Judaism is not really an '-ism' but a complete way of life, and reverence for life is its essence. There is not too much in the way of theology or dogma. The existence of God (spelled G–d by some as an indication of reverence) is real and true and basic, and the divine qualities are enumerated. There is the world, the universe, all levels and forms of reality, material and otherwise, which are God's creation. There are moral guidelines for man's behaviour in the Bible, especially the Ten Commandments, and in the Torah, the first five books of what is known outside Judaism as the Old Testament, guidelines and laws, explained and refined by rabbis and commentators over the millennia, especially in the Mishnah and Talmud, great exegetical works composed in the first 500 years of the Common Era (CE being the Jewish equivalent of AD). And this commentating, explaining and adapting has gone on to the present day.

There is a special people, the Children of Israel, the Jews, uniquely receptive to God's commands and laws. There is the land of Israel, especially dear to Jews, and to Judaism. There is an emphasis on good deeds, on action, on practical affairs, as detailed in many of the contributions you will read. There is belief in an after-life for individuals, but no great concentration on this theme, although the faith believes that 'the righteous in all nations have a share in the world to come'. There is no doctrine of original sin, but a belief in man's free will, his freedom to choose the course he will take, vital in environmental matters. Poverty is no virtue, nor is asceticism. Family life is paramount. Judaism has no monasteries or convents, monks or nuns. Holiness, whether of the Sabbath day or of the people themselves, is what matters. The word religion does not appear in the Old Testament. The word holy appears constantly.

Judaism believes that learning and study, especially of the Torah, often interpreted as the whole body of Jewish teaching, are of no less importance than prayer, not surprising for 'the people of the Book'. It believes a Messiah will usher in a Messianic Age, and therefore looks forward as well as back. Between God and man there is no intermediary. There may be inspired and learned teachers, but no human being is free from sin and uniquely divine. God, Torah, the people of Israel, the land of Israel, and the relationship between them, are at the heart of the religion. This briefly is Judaism, an uncomplicated faith, the guiding light of Jews since the patriarchs of old.

Abraham indeed has been the figure of unity between Judaism and her daughter faiths, Christianity and Islam. It is a matter of joy that those great religions have continued many of the teachings of the Jewish faith, though of course with essential and significant points of departure. Unlike them, Judaism is not now a missionary faith seeking converts, but rejoices that all men and women can come to God by their own particular route, whether Christian, Muslim, Hindu, Buddhist, Bahai, Sikh, Jain or Parsee.

Inevitably Judaism shares with Christianity and Islam common features, arising from a shared inheritance. The Christian will quote the Psalms: 'The earth is the Lord's and the fullness thereof'. The Muslim will refer to man as God's steward for creation. Hence our views as to ecology may coincide in some respects, and differ in others.

But enough of generalities. It is time to look more specifically at the tremendous wealth of Jewish teaching on ecology and the environment. These teachings lay down principles, applicable not only to bygone days, but also to our present problems of sea, land and air pollution, the greenhouse effect, holes in the ozone layer, waste disposal, and all the many ills afflicting the earth and its inhabitants today.

I hope this short work will serve as an introduction to a great theme enunciated by a great faith and may inspire you to learn more of Judaism, and of other faiths, not merely to learn, but to act, not only to study, but to perform good deeds, to rescue our environment from our own self-abuse, and to create a world that is an honour to man and a glory to his Creator.

Aubrey Rose
London

GLOSSARY

PEOPLE

Whilst there is a vast pantheon of personalities who could have been included I have chosen the main ones mentioned in the texts.

Abravanel, Isaac ben Yehuda (1437–1508) Philosopher, statesman and biblical commentator. Lived in Portugal, Spain and Italy.

Akiva, Ben Joseph (c. 40–c. 135 CE) Outstanding rabbi, scholar and teacher. Collated oral law. Executed by Romans in Caesarea.

Albo, Joseph (c. 1380–1435) Religious philosopher, author of *Sefer ha-Ikkarim* (Book of Dogmas). Lived in Spain.

Caro, Joseph (1488–1575) Religious codifer, compiled authoritative code, the Shulhan Arukh; rabbi and mystic. Lived in Spain, Turkey and Palestine.

Hertz, Joseph Herman (1872–1946) Appointed Chief Rabbi in Britain in 1913. Author of English commentaries on Pentateuch and prayer book.

Hillel, called the Elder (first century BCE) Famous Mishnaic sage and teacher. Founder of school of Hillel. Noted for his humility and tolerance. Born in Babylon, studied and taught in Jerusalem.

Hiyya, Bar Abraham (third century) Halakhist and talmudist. Born in Babylon, taught in Palestine.

Huna, or Hona (c. 216–297) Rabbi, teacher, and noted halakhist in Babylon.

Ibn Ezra, Abraham (1089–1164) Poet, biblical commentator, philosopher, scientist. Born in Spain, travelled widely, including England.

Judah Ha-Nasi (c. 135–c. 220) Patriarch of Palestinian Jewry. Often called simply Rabbi (teacher *par excellence* because of his great learning). Compiled the oral law in the Mishnah. Lived in Galilee.

Luria, Isaac (1534–72) Jewish mystic and kabbalist. Lived in Safed, Galilee.

Luzzatto, Moses Hayyim (1707–46) Poet, kabbalist, teacher. Born and lived in Italy, died in Palestine.

Maharal, Rabbi Judah Loew Ben Bezalel (c. 1525–1609) Leader and teacher in Prague, educationalist and kabbalist. Linked to legend of the Golem.

Maimonides, Moses Ben Maimon, known as Rambam (1135–1204) Rabbi, philosopher, halakhist, medical writer and physician. Main works included *Mishneh Torah* and *Guide for the Perplexed*. Author of *Thirteen Articles of Faith*. Born in Cordova, Spain, became spiritual leader in Cairo, believed buried in Tiberias.

Nahmanides, Moses ben Nahman, known as Ramban (1194–c. 1270) Rabbi, scholar, talmudist, kabbalist. Born in Spain, died in Palestine.

Nathan (second century) Rabbi in Palestine and Babylon. Treatise Avot attributed to him.

Philo (c. 20 BCE–after 40 CE) Philosopher and writer on Judaism, influenced by Greek literature and thought. Lived in Alexandria, Egypt.

Rashi, Solomon Yitzhaki (1040–1105) Noted commentator on Bible and Talmud and rabbinical authority. Lived in France.

Resh Lakish, Simeon Ben Lakish (third century) Originally a gladiator, he turned to sacred studies. Became scholar and talmudist. Lived in Palestine.

Saadiah Gaon, Saadah Ben Joseph (882–942) Leader of Babylonian Jewry, philosopher, scholar. Translated Bible into Arabic. Born in Egypt, lived in Babylon.

Shalom Aleichem (1859–1916) Pen name of Sholom Rabinovitch. Yiddish writer and humorist. Born in Russia, died in USA.

Soloveitchik, Joseph Dow (1903–   ) Talmudist and philosopher. Orthodox leader. Born in Poland. Lives in USA.

Yochanan Ben-Zakkai (first century) Rabbi, pupil of Hillel, leader of Pharisees. After Jewish revolt against Rome (66–70 CE), he founded at Yavneh a teaching centre that secured Jewish learning. Born and lived in Palestine.

Zalman, Shneour (1745–1813) Rabbi, founder of Habad Hasidism, a religious movement in Eastern Europe. Born in Lyady.

TERMS

Since Hebrew uses its own ancient alphabet, there is more than one way of writing Hebrew words in English. In this book, each author follows the system he or she prefers.

bal tashchit Thou shalt not destroy. Rabbinical teaching to prevent waste and destruction.

beracha Blessing.

CE Denotes Current Era, i.e. AD. BCE denotes Before the Current Era.

chayim (or hayim/hayyim) Life.

gemilut hasodim Personal acts of kindness as described in the Mishnah.

Gemara Commentaries on the Mishnah forming part of the Talmud.

Halacha (Halakha) Legal decisions in talmudic and rabbinical law in contradistinction to Aggada, which were non-legal homiletic expositions of the Bible and traditional teaching.

hametz Fermented dough and bread and other foodstuffs whose use is prohibited during Passover.

Jubilee The year after seven sabbatical years, i.e. the fiftieth year when, according to the Bible (Leviticus), cultivation is prohibited and special laws apply.

kashrut Biblical dietary laws, interpreted by the oral law and rabbinical teachings, as to prohibited and permitted (kosher) foods.

kiddush 'Sanctification': ceremonial blessing recited at home and in synagogue on Sabbaths and Holy Days over bread and wine.

Midrash Rabbinic interpretations or stories to illustrate significance or deeper meaning of points in Scripture, Halacha or Aggada.

Mishnah Legal codification of the oral law, compiled by Rabbi Judah Ha-Nasi, and arranged in six parts or Orders.

mitzvah Literally 'commandment': e.g. the Pentateuch contains 613 commandments or mitzvot. Widely used also for any good or kindly deed.

olim Plural of Oleh, one who 'goes up' or settles in the land of Israel.

Oneg Oneg Shabbat, literally 'Sabbath delight'. A celebration at the end of the Sabbath day.

Pesach Festival of Passover, commemorating the Exodus from Egypt, the 'season of our liberation'.

Pirke Avot Literally 'chapters of the fathers', usually known as Ethics of the Fathers, in fact part of the Mishnah containing rabbinical teachings, particularly on ethical themes, covering the third century BCE to third century CE.

Responsa Written opinions by leading rabbinical authorities responding to questions raised as to aspects of Jewish law.

Rosh Hashanah Jewish New Year, observed at the beginning of the Hebrew month of Tishri, marked by special synagogue prayers and service.

Shabbat The Sabbath, or day of rest, commencing at sunset on Friday evening and ending after nightfall on Saturday, ordained in the Ten Commandments.

shalom Hebrew for 'peace', used as a word of greeting by itself or as *shalom aleichem*, 'peace be unto you'.

Shavuot Literally 'weeks'. The name of the last of the three pilgrim festivals, linked originally to the wheat harvest, but later to celebrate the giving of the Law on Mount Sinai.

Shema The first word of the central, most significant prayer in Judaism, 'Hear, O Israel, the Lord our God, the Lord is One'.

Shemitta The seventh or Sabbatical year when, according to biblical law, the land must lie fallow and enjoy a 'solemn rest'.

shofar Horn of a ram or other permitted animal to be blown at the New Year and at other significant events.

Simchat Torah 'Rejoicing of the Law': celebration at end of Succot of completion of cycle of readings from the Torah.

Succot Literally, 'Tabernacles': pilgrim festival in month of Tishri when temporary booths are used in memory of the Children of Israel's wanderings in Sinai. Also a harvest festival. Four plant species: palm, citron, myrtle and willow, play a role in the celebration.

Talmud 'Teaching': the Babylonian and Palestinian Talmuds are vast compilations of rabbinical discussions and laws, including the Mishnah and Gemara, assembled during the centuries from 200 CE to 500 CE.

Torah Specifically the Written Torah, i.e. first five books of the Pentateuch and the oral law as handed down. More generally the term is often applied to the whole body of Jewish religious teaching.

Tosefta A collection of teachings and traditions closely related to the Mishnah.

Tu Bi Shevat Literally, 'fifteenth day' of Hebrew month of Shevat. Celebrated as the New Year for Trees when tree-planting is customary.

Yom Kippur Day of Atonement. Most solemn day of the Jewish year, spent in prayer, repentance and fasting.

ACKNOWLEDGEMENTS

The editor would like to thank the following individuals:

Chief Rabbi Dr Jonathan Sacks, Rabbi Dr Norman Solomon, Rabbi Shlomo Riskin, Rabbi Hillel Avidan, Rabbi Eli Turkel, Philip L. Pick, Vicky Joseph, Sheila Chiat, Professor Uriel N. Safriel, Professor Michael Avishai, Liat Collins, Yosef Orr, Yossi Spanier, Sammy Jackman, Clive S. M. Cohen, and Harry Blacker for the cartoons.

The editor's thanks are also extended to the following organizations:

The Jerusalem Post, Shomrei Adamah, the ULPS Rabbinic Conference, the International Jewish Vegetarian and Ecological Society, the Israel Nature Reserves Authority, Jerusalem and University Botanical Gardens, the Sternberg Centre for Judaism, the Reform Synagogues of Great Britain, the Board of Deputies of British Jews, the *Long Island Jewish World*, *The Philadelphia Inquirer*, the *Journal of Halacha and Contemporary Society*, and last but not least, the Fourth Temple Fortune Brownie Guide Pack.

Humans v Nature

All is quiet, all is still,
The trees have leaves,
The grass and hill
Do not make a noisy sound
Until the day men came around.

Black smoke so early in the day,
That all the day is turned away,
With day and night and day and night and
Nature must go by the way.

Black pillars for those beauteous trees,
Man gives nature this grave sight
Of death, destruction, and of blight.
And death and death and death and death
Of nature, goodness, and of . . . MAN.

When man is under graves of earth,
The country will be quiet and still.
The trees have leaves,
The grass and hill
Do not make a noisy sound,
When men are underneath the ground.

David Colman Rose (1957–78)
written at age 13

BLESSINGS

Blessings pervade Judaism. Hardly an event occurs that does not invoke a blessing, blessings of pleasure before eating and drinking, blessings before performing a commandment (*mitzvah*), blessings as part of thanks, blessings as part of prayer. Indeed one service includes a section known as the Eighteen Blessings.

The Hebrew word 'blessed' (*baruch*), linked to the idea of bending the knee, is as old as the very first story in the Bible. After creating man, both male and female, 'God blessed them'. Blessings flow through the subsequent stories: Abraham, Isaac, Jacob, Moses, David, Solomon, and at every significant period in Jewish history. Particularly notable are old Jacob's blessings of his many sons, and the blessings heaped on the people, according to Moses, provided they obeyed God's commandments.

Common to Judaism and Christianity is the three-fold priestly blessing of the Levites, while the Sermon on the Mount in the New Testament, continuing Jewish tradition, is full of beautiful blessings. 'Blessed is he who comes' is a daily expression in modern Hebrew conversation, as are the traditional blessings on lighting candles, on wine and bread each Sabbath Eve at home, when young children of the family are also blessed.

There is a blessing for good or bad news, for the joy of seeing a learned and wise man, on the appearance of the New Moon. When great natural events or beautiful trees or animals are witnessed, we bless God, 'who has made such as these in thy world'.

Considering the ups and downs of our long history, perhaps the blessing most favoured by Jews is the one that reads 'Blessed are thou, O Lord our God, King of the universe, who has kept us in life, and has preserved us, and has enabled us to reach this season'. The prayers which follow are but a few of many, yet so evocative of Judaism's sense of awe and thankfulness in the face of the wonders of God and nature.

Blessings in praise of life and its Creator

OVER BREAD

Blessed is the Eternal our God, Ruler of the universe, who brings forth bread from the earth.

ON TASTING FRUIT FOR THE FIRST TIME IN THE SEASON

Blessed is the Eternal our God, Ruler of the universe, for giving us life, for sustaining us, and for enabling us to reach this season.

ON SEEING BEAUTIES OF NATURE

Blessed is the Eternal our God, Ruler of the universe, whose world is filled with beauty.

ON SEEING RIVERS, SEAS, MOUNTAINS AND OTHER NATURAL WONDERS

Blessed is the Eternal our God, Ruler of the universe, who makes the wonders of creation.

ON SEEING SHOOTING STARS, ELECTRICAL STORMS AND EARTHQUAKES

Blessed is the Eternal our God, Ruler of the universe, whose power and might pervade the world.

ON SEEING TREES IN BLOSSOM

Blessed is the Eternal our God, Ruler of the universe, whose world lacks nothing needful, and who has fashioned goodly creatures and lovely trees that enchant the heart.

ON SEEING THE OCEAN

Blessed is the Eternal our God, Ruler of the universe, Maker of the great sea.

ON SEEING A RAINBOW

Blessed is the Eternal our God, Ruler of the universe, who remembers the covenant with Noah and keeps its promise faithfully with all creation.

1 | A PERSONAL VIEW

Aubrey Rose

The short piece that follows was published in The New Road, *the journal of the World Wide Fund for Nature (WWF), one of the leading world environmental bodies, presided over by the Duke of Edinburgh. His Royal Highness has time and again warned of increasing problems caused by over-population, over-development, and lack of respect for nature and the many species who jointly inhabit this world alongside us. Just as rights imply duties, so dominion implies respect and responsibility.*

Different causes impel us suddenly to become aware of problems. I have described in this article what alerted me to the growing threat to nature. There are those, regretfully, of whom it may still be said with the Psalmist, 'Eyes have they, but they see not; they have ears, but they hear not'.

I hope, as you persist with this book, your eyes and ears will open ever-wider, as mine have. Needless to say, coming from a Jewish tradition, I am still learning.

Like many others my active involvement in caring for the environment is of recent origin. Of course I was delighted, and honoured, to represent the Jewish community at the historic meeting in July 1988 when representatives of seven faiths addressed the joint Parliamentary Committee on Conservation. But the mainsprings of my present concern go back a long way.

My parents were pre-1914 immigrants to Britain from East Europe where my father, like many Jews, grew up in the country-side. As a child I was fascinated to see how he turned a few square yards of earth in the so-called garden of an East End London slum into a home for flowers, vegetables, chickens, even a vine. The memory stayed with me so that gardens and flowers have since held pride of place among my interests.

The second prompting came from a strange, but parallel source. Due to the sickness of my elder son, I studied causes of disease, and learned how slow medical officialdom was to recognize the cause of problems. From my studies, I observed that the Edinburgh College of Health had as long ago as 1910 set out the link between smoking and cancer and heart illness, yet it took 60 years before the medical profession acknowledged the fact. This set me thinking about the small voice crying in an expanding ecological wilderness, Greenpeace, Friends of the Earth, WWF bodies. Why should they not be close to a number of truths mainstream Britain persistently ignored? I began to read more and ask questions. When I asked a pioneer of the deep-litter system of chicken-rearing, packing thousands of birds into one long house, with limited natural light and air, about the likelihood of disease from the intensive and artificial feeding, watering and cleaning process, he pooh-poohed my question. The demand was there. People wanted cheap chickens, cheap eggs. That was the only way they could have them. Such was his answer. We have seen the consequences.

A third prompting was equally unusual. Inner-city riots in Britain in 1981 resulted in the setting-up of the Scarman Tribunal of Enquiry. One of my fellow-lawyers there intrigued me. He said he had scientific evidence that the amount of lead in the air in one neighbourhood was high enough to unsettle the behaviour of a group of young men, already provoked and insecure, and would lead on to violence. Not only lead, we have since learned, but other pollutants, can disturb human well-being and behaviour.

My eminent colleague however decided not to produce his evidence. He felt no one would take him seriously. How the world has changed since then!

Those of us brought up in the Jewish tradition know of the enormous respect our faith has for the land and its produce. Almost all Jewish festivals are agricultural in origin. Our prayers include a plea for rain, reflecting our origins in the Holy Land, even though we continued to live in rain-drenched countries. We know of the year of rest, Shemitta, when the land is not tilled. We know of the love of trees. The Bible is full of allusions to hills and mountains, rivers and lakes, flowers and trees. There is barely a prophet who can resist comparing the health of an individual or nation to a natural feature. The Psalmist has an intimate knowledge of flora and fauna.

This spirit was continued by the rabbis who taught 2,000 years ago. They advocated the creation of green belts and the proper disposal of waste. They enunciated a doctrine of 'thou shalt not destroy', almost a new commandment.

As a result Jews, a tiny handful of people, have been active in science, agriculture and other creative fields, although to some extent today's generation have often lost sight of their environmental heritage and need to be reminded of the importance of nature and the environment.

Many conservation schemes have been carried out in Israel, and yet mistakes have been made there, as in almost every other country of the world. People have begun to see that development is not automatically the same as progress. Draining of swamps and wetlands sounds fine, but the ecological consequences were not fully thought out.

Our religious heritage, Jewish and non-Jewish, teaches us that we are trustees of God's world for our own species and other forms of life. We, a guilty generation, have just begun to understand our crimes, as we destroy species and habitats galore. Our headlong rush into greater production, to meet unparalleled and often artificially stimulated demand, has rebounded on our heads. We are suddenly vitally concerned about the ozone layer, the quality of our air, water, food and earth. We have to pause and begin to re-learn. This is not easy.

The message of the Jewish faith was well understood by English poets like Wordsworth and William Blake. They were aware of the power, the force, behind the ostensible and the physical.

There is something spiritual that pervades all things. We do not quite understand it. We do know that it is bound up with our religion, our sense of holiness. But we are now beginning to perceive only too clearly that if we continue to destroy our environment, our environment will, in the end, destroy us.

6

SECTION A
RESOURCES IN JEWISH TEACHING AND TRADITION

2 | INTRODUCTION TO THE JEWISH FAITH

Aubrey Rose

One of the enjoyable tasks of being hailed as a spokesman for a community (how little do they know!) is being asked to address teachers. In Judaism there is no one more highly regarded than a teacher. Wealth comes and goes. Physical beauty and achievement is transient, but learning is for ever.

How delightful therefore it was for me one bleak winter's evening to spend some hours with many of London's religious education school teachers.

I spoke to them on some environmental aspects of Jewish festivals, and how to use those occasions practically to spread mutual understanding in classrooms where pupils come from many religions and traditions. The teachers, themselves of many faiths and origins, had had a long course and I was their last speaker. I tried to enliven them with a few Jewish jokes, but rapidly came to the conclusion that I was no George Burns, Danny Kaye or Jackie Mason; so I stuck to the script.

What follows is a condensed version of my talk, endeavouring to convey Judaism's joy in nature as expressed in religious festivals. I hope some of my excitement percolated through to them, and through them to their students.

When you talk about the Hindu or Christian or Muslim approach, you are talking about hundreds of millions of people, enormous populations. But when you talk about the Jewish people, you are talking about a tiny people, fewer today in the whole wide world than the population of Greater Cairo, far fewer than the populations of São Paulo or Mexico City.

9

So, unbelievable though it might seem, because we tend to get involved in so many issues, we are a small people, a small religious community, a small body yet with a great soul. And at the back of every Jewish activity and statement today, stated or unstated, is the searing fact that one-third of our people were exterminated within living memory at the hands of a brutal racism.

The second fact is that we see ourselves as part of a continuous tradition and history going back well over 3,500 years. The stories of the Old Testament are our common heritage, certainly to Christians and Muslims, but the last 2,000 years, post-biblical, have seen an incredible wealth of religious development and teaching, on the environment and so much else, as well as the development of an amazing literature, in Hebrew, in Yiddish (a form of mediaeval German infused with Hebrew) and in Ladino, a Spanish-based tongue also full of Hebrew, found among the Sephardi or Mediterranean Jews as opposed to the Ashkenazim, the Central and East European Jews. And that literature itself is quite remarkable. You may have heard of all the American Jews writing in English: Herman Wouk, Saul Bellow, Norman Mailer, Bernard Malamud, or the host of Jewish writers in Britain: Bernard Levin, Brian Glanville, Harold Pinter, Peter Shaffer, etc., but in the last twenty years the Nobel prize for literature has in fact been won by two writers who wrote in Hebrew and Yiddish, great authors, Shmuel Agnon and Isaac Bashevis Singer.

We have been the people of the Book. I wonder whether we will become the people of the computer and the word-processor?

Two of our most famous books are the Mishnah and the Talmud; the latter actually consists of 63 books. The Mishnah was compiled about 200, summarizing biblical laws, whilst the Talmud, a compilation of rabbinical commentaries and views, grew during the next 300 years, enlarging on the basic laws. The Talmud contains over three million words, and took five centuries to complete, with contributions from over 2,000 authors. It is the most democratic book ever published. No opinion, however diverse or contrary, is ever suppressed. While Judaism is monotheistic, it is not monolithic. There may be one God, but there is a profusion of competing opinions. The Talmud concerns every phase of human activity. In it you will find laws about retaining green belts around cities, about waste disposal, and the doctrine of *bal tashchit*—you shall not destroy. You are on this earth to preserve and guard nature, not to destroy it.

This follows the biblical injunction to soldiers in battle not to destroy trees, especially fruit trees. Maimonides, the great mediaeval Jewish doctor, author, and religious rationalist, who incidentally wrote some of his greatest works in Arabic, declared: 'It is not only forbidden to destroy fruit-bearing trees, but whoever breaks vessels, tears clothes, demolishes a building, stops up a fountain or wastes food in a destructive way offends against the law of "do not destroy"'.

The rabbis take the doctrine further. Don't waste your wealth or your talents. Don't take drugs because they damage the mind and destroy your potential as a human being. Don't punish the body by self-inflicted wounds. Don't destroy animals unless you have to for food or self-preservation. Certainly don't hunt them for pleasure. Do always preserve nature. Do not destroy it. It is God's gift to us.

The Old Testament is full of the wonders of nature, of knowledge of nature. The Book of Psalms is a collective hymn to God and to nature. The prophet Isaiah raised the description of the land, rivers, trees, flowers, animals, to a sublime level, in imperishable words and visions of peace. Well over 2,500 years ago a Jewish prophet talked about the day when men would learn war no more. We are still light-years away from achieving that vision, but it must always remain our vision.

If you have been to a Jewish religious service, you will hear three words time and again, *chayim*, meaning life, *Shabbat* meaning Sabbath, and *shalom* meaning peace. These words summarize the Jewish approach to our environment.

Chayim, life, is at the heart of traditional Jewish teaching. We say *le chayim*, to life, when we raise our glass of wine. Israel has no death penalty even where life has been taken: only one person has been executed since 1948, Adolf Eichmann, under a special law. We say in our prayers 'Cause us, O Lord, to lie down in peace, and raise us up, O our King, to enjoy life'. Life is to be enjoyed, abundantly, as Jesus, from his Jewish tradition, taught.

The Sabbath, Shabbat, is a uniquely Jewish institution, when everyone rests. This includes people, and animals, and the land itself. The environment needs rest too.

Shalom, peace, is linked to the Sabbath. In traditional Judaism the land was not to be worked at all every seventh year, but to lie untouched and at peace. It is a doctrine of respect for all forms of life, recognizing their interdependence, almost with overtones of the modern theory of Gaia, the self-regulating systems of the earth.

These then are the basic elements behind Jewish teaching. Later on I quote some of the wonderful illustrations of these teachings, and you may also wish to read that wonderful love poem, the Song of Songs, reflecting great knowledge of trees, plants, flowers and animals, and all living things.

We have a whole series of fasts, the best-known being Yom Kippur, the Day of Atonement. We have a fast known as Tisha Be'av, commemorating the destruction of the Temple in Jerusalem in AD 70. And we have added new days to our religious calendar, not merely Israel Independence day—Israel incidentally, became independent before more than two-thirds of the present members of the UN—but we also have a Yom Hashoah, a Holocaust Memorial Day. There are new special days, but the core of the calendar are the festivals, dating from biblical times. We have no shortage of festivals. The New Year festival (Rosh Hashanah) is solemn, the day of Judgement, of Memorial, of sounding the shofar, beginning a period of self-examination known as the Ten Days of Penitence. It is a day we like to describe as the birthday of the world.

Then there are the three pilgrim festivals, when traditionally all Jews took to the road, journeying up to Jerusalem, much as the Muslim makes the Haj, or the Hindu travels long distances to wade in the Ganges at Benares. These festivals are Passover (Pesach), commemorating the Exodus from Egypt—the season of our liberation—Shavuot, the Feast of Weeks or Pentecost, recalling the giving of the Law on Mount Sinai, also a festival of First Fruits and of Harvest, and lastly Succot (Tabernacles) reminding us of the temporary structures in which the Israelites lived in the desert, as temporary as our lives, but also a time of rejoicing and of harvest.

Never in these three festivals are we far away from nature. The latter, Tabernacles, ends with a riotous celebration called Simchat Torah when we all dance around the synagogue holding the scrolls of the Law, '. . . and David danced'.

In addition there is Chanukah, usually around Christmas time, a Festival of Lights in mid-winter, but much more than just that. There is the New Year for Trees (Tu Bi Shevat), a special day devoted to the planting of trees, usually in February, and of course Purim, the story of Queen Esther in Persia, the nearest thing to a Jewish Carnival, when we are involved in parades, fancy dress and presents, a great fun festival.

We are not a drinking people. We have hardly any drunkenness in our community, here or abroad. Maybe it's because we are forced to drink four glasses of wine in the Passover service, and a glass of wine every Friday night at home. An obligation to drink wine removes any of the secret glamour of drinking, and becomes as exciting an exercise as eating a piece of bread.

In the musical *Fiddler on the Roof*, Tevye, a poor peasant farmer, lives in a little village, with his cow and chickens, and the horse to whom he talks. Many Jews lived like him on the land. My father did, in East Europe. When they fled from persecution they settled in big cities, London, New York, just as the West Indians in Britain settled in the cities, although most of them also come from rural origins and backgrounds.

The creator of the story of Tevye is a great Yiddish humorist known as Shalom Aleichem. He knew his Jews well when he said that at the festival of Purim Jews get all excited about having an extra drink; in fact, as he put it, they get drunk on the very thought of getting drunk.

These festivals have several things in common. None of them commemorates an event in the life of an *individual*, like Christmas, or the birthday of the Prophet. They are all to do with special events in the life of a *people* and a land. Passover relates to the Exodus from slavery in Egypt, Pentecost to the receiving of the Law, Succot or Tabernacles to wandering through Sinai. They are all festivals of the people by the people, and for the people. Individuals take second place.

Another feature is that they all relate to the Land of Israel, to harvests, trees, planting, to wheat and barley. They began as agricultural festivals of thanksgiving, and developed special characteristics linked to the ethical and national history of the people.

Jews have a passion for food, with some of our rabbis, like Lionel Blue, writing cookery books. Each festival has mouth-watering special foods eaten, for example, at Purim, or at Pentecost when the Torah or Law was given. Often the food is symbolic, but always a great treat, often depending on the part of the world the Jews lived in. We eat an apple dipped in honey at New Year, seeking a sweet year ahead. At Passover the table is full of symbolic foods, especially the Matzo or unleavened bread.

Perhaps the best environmental link is Succot, Tabernacles. It occurs at the time of the last fruit harvest, a time in autumn of rejoicing and thanksgiving. Children at school make model booths or huts out of shoe boxes, decorating them with fruit and leaves and flowers. At home a temporary structure is made, similarly adorned, in which people eat their meals. From the branches that form the roof hang every kind of fruit. It is great fun to make, like building a verdant tent in the garden. In talmudic times people used to sleep in the succahs they built; some people still do. It reminds one of the hard times in the desert. In the synagogue and at home palm branches are waved during the prayers, myrtle and willow twigs and a citron are used in the ritual to the age-old cry *Hoshanna*, 'Save us, O God'. This ceremony of waving palms in all directions also echoes part of the Christian story. It links a moral concept with the world of nature.

The whole of this Tabernacle festival is environmentally linked. Yet on a deeper level it teaches that we are all wanderers here, all equal, all transient, all part of the natural world around us, a little lower than the angels, yet still also creatures of dust and ashes, whose bodies return to Mother Earth.

The whole of our tradition is linked to trees and to water, in fact and in allegory. Our story begins with the tree of knowledge of good and evil, the tree of life, right up to present-day Israel, for you cannot separate that land from the Jewish story. From the beginning of this century, under the auspices of the Jewish National Fund most Jewish homes world-wide have had little blue and white boxes in which children and parents have put a penny, a cent or a centime weekly. That money, a self-imposed community charge, has been collected and used for 90 years, to plant a tree, clear a swamp, buy a field, all geared to the Land. In one of the greatest tree-planting operations in history, over 180 million trees have been planted in the last 40 years, creating shade, preventing soil erosion, helping the water level; an operation of ecological love.

It is not surprising therefore that children love the minor festival we call the New Year for Trees. In Hebrew it is known as Tu Bi Shevat, the fifteenth day of the month of Shevat, when the winter ends in Israel. It began as a period for computing agricultural tithes, but is today purely a nature festival. Trees are planted in Israel and in Jewish communities world-wide. In the schools young children plant citrus pips in yoghourt cartons and watch with joy as the plant grows. My greenhouse is full of little citrus plants that have graduated from carton to pots.

You will find in collections of Israel's postage stamps whole series devoted to trees, or flowers, or nature, each with a biblical quotation attached. These collections can be a useful introduction to nature in the Bible. It is not at all surprising that many of Jesus' most vivid stories were agricultural in allusion, from the mustard seed to the vineyard, the good soil needed for fruitful trees to the lilies of the field. All the rabbis in those days spoke in parables.

Out of this profound regard for nature came a host of modern organizations: the Society for the Preservation of Nature in Israel, the Biblical and Botanical Gardens in Jerusalem, the Nature Reserves Authority, the extraordinary agricultural kibbutzim or settlements where equality reigns and money is of limited significance.

I have mentioned the overwhelming importance of trees, but water is another vital symbol and reality. We speak of *mayim chayim*, living waters. In a dry land wells were of great importance, as described in the stories of Abraham and Isaac. When something good happened to the patriarchs they either built an altar or dug a well. Later the Psalmist spoke of our exile 'by the waters of Babylon'. Today water is as important as ever, especially as new immigrants (*olim*) go up to the Land.

To show the importance of that Land in the consciousness of Jews, we have prayed for the past few thousands years for rain, special prayers for rain even when we lived in the most rain-soaked parts of the world, because this perpetuated the original prayers for rain in Judea and in the Land of Israel.

In line with this, the modern state has a series of laws concerning the cleanliness of the air and the water, and has set up a special Ministry of the Environment whereby every planning matter must have an environmental assessment, akin to the European Community Directive of July 1988, but much more far-reaching.

Israel also plays her role in the UN Environmental Programme seeking a cleaner Mediterranean, sitting down with nations who have no political link with her at all. The environmental cause overlaps boundaries and hostilities and political issues and forces a common heritage concept on us squabbling humans.

Thus trees, and land, and water play an important role in Jewish life, deriving from the teachings of the Bible and the rabbis.

I have referred to single words that are illustrative of the Jewish faith and the environment. Let me add another word: light. Light, according to Genesis, was the first thing God created. Without light there could be no nature, no environment. And the creation of light has ever been a Jewish theme. The absence of light was one of the ten plagues. Gideon used light in his battles. There is even the story of the sun and the moon standing still to help Joshua. The Jewish calendar is a lunar calendar. A joyous festival is Chanukah, in mid-winter, the festival of lights. Children love to make the *chanukiah*, the eight-branched candlestick, and to light an extra candle each night. It is also a time when children receive presents.

Another word, *beracha* or blessing. Our prayers are full of blessings, including those in relation to the environment. We have a blessing over the fruit of the vine, over bringing forth bread from the earth, a blessing on grapes and figs and dates, a blessing on vegetables: 'Blessed art thou, Oh Lord our God, King of the Universe, who createst the fruit of the earth', a blessing on smelling fragrant bark of a tree or sweet-smelling plants. We even have special blessings to be said on the happening of natural events, lightning, thunder, on seeing the sea or beautiful trees or animals, on seeing a rainbow, on seeing the first blossom on a tree.

And these blessings hallow nature, and respect the environment, the work of the Creator.

The festivals are celebrations of nature, all of them. They teach concern for nature and love of nature. But as in everything in Judaism, the moral teachings come through all the time. When you reap your field, leave the corners for the poor. Leave some grapes on your vines for the poor. You have an obligation to help those less well off—the stranger, the widow and the orphan. Through agriculture you teach a sense of responsibility and community.

And performing God's commandments will lead to fruitfulness.

> If you hearken diligently to my commandments then I will send rain for your land in its due season, and I will give grass in thy field for thy cattle. But if you turn aside and serve other gods, then the land will not yield her products and you will perish quickly from off the good land which the Lord giveth unto you.

There are many quotations, so many. The Psalmist understood and loved nature when he wrote

> Let the field exult, and all that is therein.
> Then shall all the trees of the wood sing for joy.

Let the words of a noted present-day ecologist, Dr David Bellamy, confirm this heritage. He writes,

> It is indeed a sobering thought that the early writings of the Jewish people encompass all the basic recommendations of World Conservation Strategy.

Sobering indeed that the lessons of long, long ago remain yet unlearned.

Every faith has much to offer, not only Judaism. None of them faced in the days of their birth the problems we face today, but each of them enunciated principles still relevant to present times. Each has a sense of holiness and above all concern for man and his world because each teaches that there is purpose in the life of each person and purpose in the world; that we are inheritors from the past and trustees for the future.

3 | JUDAISM AND THE ENVIRONMENT

Norman Solomon

I have had the privilege of working with Rabbi Dr Norman Solomon, whose contribution follows, for several years. He has been the rabbi of a prominent London synagogue, an activist in Jewish–Christian relations, and editor of a journal on that theme, published by the Institute of Jewish Affairs in London. He is currently Director of The Centre for the Study of Judaism and Jewish–Christian Relations at Selly Oak Colleges near Birmingham.

He is a man who likes to get things done, and as my Vice-Chairman on the Board of Deputies Working Group on the Environment he has frequently urged me on with a challenging 'Nu', an untranslatable Yiddishism. Highly respected and learned, particularly for his leading role in interfaith work in Britain and abroad, his essay is of especial significance, drawing deeply on biblical and rabbinical sources, going to the heart of the theoretical and practical approach of Judaism to conservation and the environment.

When the religion has something to say, he says it. When it has no contribution to make, he says so. His yea is a yea, and his nay is a nay. He deals honestly with problems, whether on the subject of hunting, nuclear energy or possible conflict in biblical texts.

What Rabbi Solomon does show, in his closely reasoned contribution, based on his personal study and assessments, and marked by scholarship and apt quotation, is the deep reverence of Judaism for human life, of concern for other forms of life, of the need for religions to co-operate rather than confront one another. His is the practical, sensible approach, typical of the modest man he is and the sincere faith he espouses.

PSALM 148

1   Praise the Lord!
    Praise the Lord from the heavens;
        praise him in the heights!
2   Praise him, all his angels;
        praise him, all his host!
3   Praise him, sun and moon;
        praise him, all you shining stars!
4   Praise him, you highest heavens,
        and you waters above the heavens!
5   Let them praise the name of the Lord,
        for he commanded and they were created.
6   He established them forever and ever;
        he fixed their bounds, which cannot be passed.
7   Praise the Lord from the earth,
        you sea monsters and all deeps,
8   fire and hail, snow and frost,
        stormy wind fulfilling his command!
9   Mountains and all hills,
        fruit trees and all cedars!
10 Wild animals and all cattle,
        creeping things and flying birds!
11 Kings of the earth and all peoples,
        princes and all rulers of the earth!
12 Young men and women alike,
        old and young together!
13 Let them praise the name of the Lord,
        for his name alone is exalted,
        his glory is above earth and heaven.
14 He has raised up a horn for his people,
        praise for all his faithful,
        for the people of Israel who are close to him.
    Praise the Lord!

A Jewish statement on nature

1 Creation is good; it reflects the glory of its creator.
'God saw everything he had made, and indeed it was very good'
(Genesis 1:31). Judaism affirms life, and with it the creation as a whole.

2 Biodiversity, the rich variety of nature, is to be cherished.
In Genesis 1, everything is said to be created 'according to its kind'. In
the story of the flood, Noah has to conserve in the ark male and female
of every species of animal, so that they may subsequently procreate.

3 Living things range from lower to higher, with humankind at the top.
Genesis 1 depicts a process of creation of order out of the primaeval
chaos. The web of life encompasses all, but human beings—both male
and female (Genesis 1:27), 'in the image of God'—stand at the apex of
this structure.

4 Human beings are responsible for the active maintenance of all life.
Setting people at the top of the hierarchy of creation places them in a
special position of responsibility towards nature. Adam is placed in the
garden of Eden 'to till it and to preserve it' (Genesis 2:15), and to 'name'
(that implies, understand) the animals.

5 Land and people depend on each other.
The Bible is the story of the chosen people and the chosen land. The
prosperity of the land depends on the people's obedience to God's
covenant: 'If you pay heed to the commandments which I give you this
day, and love the Lord your God and serve him with all your heart and
soul, then I will send rain for your land in season . . .' (Deuteronomy
11:13–17). In the contemporary global situation, this means that
conservation of the planet depends on (a) the social justice and moral
integrity of its people and (b) a caring, even loving, attitude to land,
with effective regulation of its use.

6 Respect creation—do not waste or destroy.
Bal tashchit ('not to destroy'—see Deuteronomy 20:19) is the Hebrew
phrase on which the rabbis base the call to respect and conserve all that
has been created.

21

Judaism and the environment[1]

1 INTRODUCTION

It is widely recognized today that people are destroying the environment on which living things depend for their existence. Many species are endangered as a result of human activity, the planetary climate may already have been destabilized, the protective ozone layer has been damaged, forests have been destroyed, species threatened or made extinct, and pollution in forms such as acid rain and other forms of water contamination is widespread.

Much of this destruction arises from the level of economic activity demanded by a rapidly increasing world population which is locally raising its living standards faster than ecologically sustainable levels of production.

In addition, there is a permanent worry that stockpiles of highly destructive weapons might actually be used, and that the use of even a small part of the available arsenal would cause irreversible damage to the planetary environment, perhaps rendering impossible the survival of humanity and many other species.

It is not at first sight clear what these problems have to do with religious beliefs. After all, the only belief necessary to motivate a constructive response to them is a belief in the desirability of human survival, wedded to the perception that human survival depends on the whole interlinking system of nature. The belief is not peculiar to religions, but part of the innate self-preservation mechanism of humankind; the perception of the interdependence of natural things arises not from religion, but from careful scientific investigation.

Moreover, the discovery of which procedures would effectively solve the problems of conservation is a technical, not a religious one. If scientists are able to offer alternative procedures of the same or different efficiency the religious may feel that the ethical or spiritual values they espouse should determine the choice. But few choices depend on value judgements alone, and no judgement is helpful which is not based on the best available scientific information.

22

These considerations will be borne in mind as we examine the relevance of traditional Jewish sources to our theme.

In the following pages we offer a structured guide to the main traditional Jewish sources which relate to the great environmental problems of our time. Judaism did not 'stop' with the Bible or the Talmud; it is a living religion constantly developing in response to changing social realities and intellectual perceptions. At the present time, it is passing through one of its most creative phases; however, within the limited scope of this article only a few references can be made to the contemporary literature.

Traditional Jewish thought is expressed in several complementary genres. The most distinctive is halakha, or law, but history, myth, poetry, philosophy and other forms of expression are also significant. Our focus here is not on the contributions made by individual Jews, for instance scientists and economists, to the modern ecological movement—this would make an interesting study in itself—but on the religious sources, which demonstrate the continuity between traditional Jewish thought and a range of contemporary approaches.

2 ATTITUDES TO CREATION

2.1 Goodness of the physical world

'God saw that it was good' is the refrain of the first creation story of Genesis (chapter 1:1 to 2:4), which includes the physical creation of humankind, male and female. The created world is thus testimony to God's goodness and greatness (see Psalms 8, 104, 148, and Job 36:22 to 41:34).

The second 'creation' story (Genesis 2:5 to 3:24) accounts for the psychological makeup of humankind. There is no devil, only a 'wily serpent', and the excuse of being misled by the serpent does not exempt Adam and Eve from personal responsibility for what they have done. Bad gets into the world through the free exercise of choice by people, not in the process of creation, certainly not through fallen angels, devils, or any other external projection of

23

human guilt; such creatures are notably absent from the catalogue of creation in Genesis 1.

Post-biblical Judaism did not adopt the concept of 'the devil'. In the Middle Ages, however, the dualism of body and spirit prevailed, and with it a tendency to denigrate 'this world' and 'material things'. The Palestinian kabbalist Isaac Luria (1534–72) taught that God initiated the process of creation by 'withdrawing' himself from the infinite space he occupied; this theory stresses the 'inferiority' and distance from God of material creation, but compensates by drawing attention to the divine element concealed in all things. The modern Jewish theologian who wishes to emphasize the inherent goodness of God's creation has not only the resources of the Hebrew scriptures on which to draw but a continuous tradition based on them.

The Bible encompasses three realms, of God, of humankind, of nature. It does not confuse them. Its 'creation spirituality' articulates 'original blessing' rather than 'original sin'—'God saw all that he had made, and it was very good' (Genesis 1:31)—and this includes all creatures, culminating with humans. As Aaron Lichtenstein remarked at a conference on Judaism and Ecology at Bar Ilan University (Tel Aviv), 'Our approach is decidedly anthropocentric, and that is nothing to be ashamed of'.[2]

Below, in section 2.4, we shall speak of the hierarchy within nature itself.

2.2 Biodiversity

I recall sitting in the synagogue as a child and listening to the reading of Genesis. I was puzzled by the Hebrew word *leminehu*, 'according to its kind', which followed the names of most of the created items and was apparently superfluous. Obviously, if God created fruit with seeds, the seeds were 'according to its kind'!

As time went on I became more puzzled. Scripture seemed obsessive about 'kinds' (species). There were careful lists and definitions of which species of creature might or might not be eaten (Leviticus 11 and Deuteronomy 14). Wool and linen were not to be mixed in a garment (Leviticus 19:19; Deuteronomy 22:11), ox and ass were not to plough together (Deuteronomy 22:10), fields

(Leviticus 19:19) and vineyards (Deuteronomy 22:9) were not to be sown with mixed seeds or animals cross-bred (Leviticus 19:19) and, following the rabbinic interpretation of a thrice repeated biblical phrase (Exodus 23:19; 34:26; Deuteronomy 14:21), meat and milk were not to be cooked or eaten together.[3]

The story of Noah's Ark manifests anxiety that all species should be conserved, irrespective of their usefulness to human-kind—Noah is instructed to take into his Ark viable (according to the thought of the time) populations of both 'clean' and 'unclean' animals. That is why the 'Inter-stellar Ark' is the model, amongst those concerned with such things, for gigantic spaceships to carry total balanced communities of living things through the galaxy for survival or colonization.[4]

The biblical preoccupation with species and with keeping them distinct can now be read as a way of declaring the 'rightness' of God's pattern for creation and of calling on humankind not only not to interfere with it, but to cherish biodiversity by conserving species.

Scripture does not of course take account of the evolution of species, with its postulates of (a) the alteration of species over time and (b) the extinction (long before the evolution of humans) of most species which have so far appeared on earth.[5] Yet at the very least these Hebrew texts assign unique value to each species as it now is within the context of the present order of creation; this is sufficient to give a religious dimension, within Judaism, to the call to conserve species.

2.2.1 Pereq shira

Pereq shira[6] (the 'Chapter of Song') affords a remarkable demon-stration of the traditional Jewish attitude to nature and its species. No one knows who composed this 'song', though it may have originated amongst the Hekhalot mystics of the fourth or fifth centuries. More significant than its origin is its actual use in private devotion. It has been associated with the 'Songs of Unity' com-posed by the German pietists of the twelfth century who undoubt-edly stimulated its popularity. At some stage copyists prefaced to it exhortatory sayings which were erroneously attributed to tal-mudic rabbis, for instance: 'Rabbi Eliezer the Great declared that whoever says *Pereq shira* in this world will acquire the right to say it the world to come'.[7]

As the work is printed today it is divided into five or six sections, corresponding to the physical creation (this includes heaven and hell, Leviathan and other sea creatures), plants and trees, creeping things, birds, and land animals (in some versions the latter section is sub-divided). Each section consists of from ten to 25 biblical verses, each interpreted as the song or saying of some part of creation or of some individual creature. The cock, in the fourth section, is given 'seven voices', and its function in the poem is to link the earthly song, in which all nature praises God, with the heavenly song.

We shall see in section 2.4 that the philosopher Albo (1380–1435) draws on *Pereq shira* to express the relationship between the human and the animal; yet *Pereq shira* itself draws all creation, even the inanimate, even heaven and hell themselves, into the relationship, expressing a fullness which derives only from the rich diversity of things, and which readily translates into the modern concept of biodiversity.

2.3 *Stewardship or domination?*

There has been discussion amongst Christian theologians as to whether the opening chapters of Genesis call on humans to act as stewards, guardians of creation, or to dominate and exploit the created world. There is little debate on this point amongst Jewish theologians,[8] to whom it has always been obvious that when Genesis states that Adam was placed in the garden 'to till it and to care for it' (2:15) it means just what it says. As Rav Kook (1865–1935)[9] put it:

> No rational person can doubt that the Torah, when it commands people to 'rule over the fishes of the sea and the birds of the sky and all living things that move on the earth' does not have in mind a cruel ruler who exploits his people and servants for his own will and desires—God forbid that such a detestable law of slavery [be attributed to God] who 'is good to all and his tender care rests upon all his creatures' (Psalm 145:9) and 'the world is built on tender mercy' (Psalm 89:3).[10] [11]

In the twelfth century the great Jewish Bible scholar Abraham ibn Ezra commented as follows on the words of Psalm 115:16, 'The heavens are the heavens of the Lord, and he gave the earth to people':

> The ignorant have compared man's rule over the earth with God's rule over the heavens. This is not right, for God rules over every-thing. The meaning of 'he gave it to people' is that man is God's steward (*paqid*—officer or official with special responsibility for a specific task) over the earth, and must do everything according to God's word.

So perverse is it to understand 'and rule over it' (Genesis 1:28)—let alone Psalm 8—as meaning 'exploit and destroy' (is that what people think of their rulers?) that many Christians take such in-terpretations as a deliberate attempt to besmirch Christianity and not a few Jews have read the discussions as an attempt to 'blame the Jews' for yet another disaster in Christendom. The context of Genesis 1:28 is indeed that of humans being made in the image of God, the beneficent creator of good things; its meaning is there-fore very precise, that humans, being in the image of God, are summoned to share in his creative work, and to do all in their power to sustain creation.

2.4 Hierarchy in creation

'God created humans[12] in his image . . . male and female he created them' (Genesis 1:27). In some sense, humankind is superior to animals, animals to plants, plants to the inanimate. There is a hierarchy in created things.

The hierarchical model has two practical consequences. First, as we have seen, is that of responsibility of the higher for the lower, traditionally expressed as 'rule', latterly as 'stewardship'. The second is that, in a competitive situation, the higher has priority over the lower. Humans have priority over dogs so that, for instance, it is wrong for a man to risk his life to save that of a dog though right, in many circumstances, for him to risk his life to save that of another human. Contemporary dilemmas arising from this are described in section 5.1.

27

The Spanish Jewish philosopher Joseph Albo (1380–1435) places humans at the top of the earthly hierarchy, and discerns in this the possibility for humans to receive God's Revelation.[13] This is just a mediaeval way of saying what we have remarked. God's Revelation, *pace* Albo and Jewish tradition, is the Torah from which we learn our responsibilities to each other and to the rest of creation.

According to Albo, just as clothes are an integral part of the animal, but external to people, who have to make clothes for themselves, so are specific ethical impulses integral to the behaviour of particular animals, and we should learn from their behaviour. 'Who teaches us from the beasts of the earth, and imparts wisdom to us through the birds of the sky' (Job 35:11)—as the Talmud puts it: 'Rabbi Yohanan said, If these things were not commanded in the Torah, we could learn modesty from the cat, the ant would preach against robbery, and the dove against incest'.[14] The superiority of humans lies in their unique combination of freedom to choose and the intelligence to judge, without which the divine Revelation would have no application. Being in this sense 'higher' than other creatures, humans must be humble towards all. Albo, in citing these passages and commending the reading of *Pereq shira* (see section 2.2.1), articulates the attitude of humble stewardship towards creation which characterizes rabbinic Judaism.

Rav Kook, drawing on a range of classical Jewish sources from Psalm 148 to Lurianic mysticism, beautifully acknowledges the divine significance of all things (the immanence of God):

> I recall that with God's grace in the year 5665 [1904/5] I visited Jaffa in the Holy Land, and went to pay my respects to its Chief Rabbi [Rav Kook]. He received me warmly . . . and after the afternoon prayer I accompanied him as he went out into the fields, as was his wont, to concentrate his thoughts. As we were walking I plucked some flower or plant; he trembled, and quietly told me that he always took great care not to pluck, unless it were for some benefit, anything that could grow, for there was no plant below that did not have its guardian[15] above. Everything that grew said something, every stone whispered some secret, all creation sang . . .[16]

2.5 *Concern for animals*

Kindness to animals is a motivating factor for general concern with the environment, rather than itself an element in conservation.

It features prominently in the Jewish tradition. The Ten Commandments include domestic animals in the Sabbath rest, and the 'seven Noahide laws' are even more explicit.[17] Pious tales and folklore exemplify this attitude, as in the talmudic anecdote of Rabbi Judah the Patriarch's contrition over having sent a calf to the slaughter,[18] and the folk tale of the Rabbi and the Frog.[19]

2.5.1 Causing pain or distress to animals

In rabbinic law this concern condenses into the concept of *tsaar baalei hayyim* ('distress to living creatures').[20] An illuminating instance of halakhic concern for animal welfare is the rule attributed to the third-century Babylonian Rav that one should feed one's cattle before breaking bread oneself;[21] even the Sabbath laws are relaxed somewhat to enable rescue of injured animals or milking of cows to ease their distress. Recently, concern has been expressed about intensive animal husbandry including battery chicken production.[22] Likewise, many rabbinic Responsa have been published on the restraints to be placed on experimentation on animals; clearly, experimentation is not allowed for frivolous purposes, but it is necessary to define both the human benefits which might justify animal experimentation and the safeguards necessary to avoid unnecessary suffering to animals.[23]

2.5.2 Meat eating

The Torah does not enjoin vegetarianism, though Adam and Eve were vegetarian (Genesis 1:29). Restrictions on meat eating perhaps indicate reservations; Joseph Albo wrote that the first people were forbidden to eat meat because of the cruelty involved in killing animals.[24] Isaac Abravanel (1437–1508) endorsed this,[25] and also taught that when the Messiah comes we would return to the ideal, vegetarian state.[26] Today the popular trend to vegetarianism has won many Jewish adherents though little official backing from religious leaders.[27]

Judah Tiktin[28] cites the kabbalist Isaac Luria (1534–72) as saying 'Happy are they who are able to abstain from eating meat and drinking wine throughout the week'. This has been cited as support for vegetarianism, but is irrelevant. The context is that of abstaining from meat and wine on Mondays and Thursdays (the traditional penitential days), a custom akin to the widespread Roman Catholic practice on not eating meat on Fridays. The goal is self-denial, or asceticism, not vegetarianism, as may be inferred from the fact that the very same authorities endorse the eating of meat and the drinking of wine in moderation as the appropriate way to celebrate the Sabbath and festivals. There have, indeed, been some holy men whose asceticism has led them to abstain entirely from eating meat and drinking wine—Rabbi Joseph Kahaneman (1888–1969)[29] for instance—but this provides no basis in principle for vegetarianism.

2.5.3 Hunting

On 23 February 1716 Duke Christian of Sachsen Weissenfels celebrated his fifty-third birthday with a great hunting party. History would have passed by the Duke as well as the occasion had not J. S. Bach honoured them with his 'Hunting Cantata'. The text by Salomo Franck, secretary of the upper consistory of Weimar, is a grand celebration of nature and its priest, Duke Christian, with no sense that hunting sounds a discordant note, and the cantata includes one of Bach's most expressive arias, *Schafe können sicher weiden* ('Sheep may safely graze').

Hunting, it is clear, enhances appreciation of nature. Moreover, the hunter does not oppose conservation; he destroys only individual prey but has an interest in preserving the species.

Conditions of Jewish life in the past millennium or so have rarely afforded Jewish princes the opportunity to celebrate their birthdays by hunting parties. But it has happened from time to time, and led rabbis to voice their censure.

Professor Nahum Rakover,[30] Deputy Attorney General of Israel, sums up the halakhic objections to 'sport' hunting under eight heads:

1 It is destructive/wasteful (see section 3.2).

2 It causes distress to animals (section 2.5.1).

3 It actively produces non-kosher carcasses.[31]

4 It leads to trading non-kosher commodities.

5 The hunter exposes himself to danger unnecessarily.

6 It wastes time.[32]

7 The hunt is a 'seat of the scornful' (Psalm 1:1).[33]

8 'Thou shalt not conform to their institutions' (Leviticus 18:3).[34]

From this we see that although Jewish religious tradition despises hunting for sport this is on ethical and ritual grounds rather than in the interest of conservation.[35]

3 THE LAND AND THE PEOPLE—A PARADIGM

Judaism, whilst attentive to the universal significance of its essential teachings, has developed within a specific context of peoplehood. In the Bible itself the most obvious feature of this is the stress on the chosen people and the chosen land.

This has meant that Judaism, both in biblical times and subsequently, has emphasized the inter-relationship of people and land, the idea that the prosperity of the land depends on the people's obedience to God's covenant. For instance:

> If you pay heed to the commandments which I give you this day, and love the Lord your God and serve him with all your heart and soul, then I will send rain for your land in season ... and you will gather your corn and new wine and oil, and I will provide pasture ... you shall eat your fill. Take good care not to be led astray in your hearts nor to turn aside and serve other gods ... or the Lord will become angry with you; he will shut up the skies and there will be no rain, your ground will not yield its harvest, and you will soon vanish from the rich land which the Lord is giving you (Deuteronomy 11:13–17).

Two steps are necessary to apply this link between morality and prosperity to the contemporary situation:

1 The chosen land and people must be understood as the prototype of (a) all actual individual geographical nations (including, of course, Israel) in their relationships with land and of (b)

31

humanity as a whole in its relationship with the planet as a whole.

2 There must be satisfactory clarification of the meaning of 'obedience to God' as the human side of the covenant to ensure that 'the land will be blessed'. The Bible certainly has in mind justice and moral rectitude, but in spelling out 'the commandments of God' it includes specific prescriptions which directly regulate care of the land and celebration of its produce; some of these are discussed below.

To sum up—the Bible stresses the intimate relationship between people and land. The prosperity of land depends on (a) the social justice and moral integrity of the people on it and (b) a caring, even loving, attitude to land with effective regulation of its use. Conservation demands the extrapolation of these principles from ancient or idealized Israel to the contemporary global situation; this calls for education in social values together with scientific investigation of the effects of our activities on nature.

3.1 Sabbatical year and Jubilee

> When you enter the land which I give you, the land shall keep Sabbaths to the Lord. For six years you may sow your fields and for six years prune your vineyards but in the seventh year the land shall keep a Sabbath of sacred rest, a Sabbath to the Lord. You shall not sow your field nor prune your vineyard ... (Leviticus 25:2–4).

The analogy between the Sabbath (literally, 'rest day') of the land and that of people communicates the idea that land must 'rest' to be refreshed and regain its productive vigour. In contemporary terms, land resources must be conserved by avoiding over-use.

The Bible pointedly links this to social justice. Just as land must not be exploited so slaves must go free after six years of bondage or in the Jubilee (fiftieth) year, and the sabbatical year (in Hebrew *shemitta*—'release') cancels private debts, thus preventing exploitation of the individual.

The consequence of disobedience is destruction of the land, which God so cares for that he will heal it in the absence of its unfaithful inhabitants:

If in spite of this you do not listen to me and still defy me . . . I will make your cities desolate and destroy your sanctuaries . . . your land shall be desolate and your cities heaps of rubble. Then, all the time that it lies desolate, while you are in exile in the land of your enemies, your land shall enjoy its sabbaths to the full . . . (Leviticus 26:27–35).

If in Israel today there is only a handful of agricultural collectives which observe the 'Sabbath of land' in its biblical and rabbinic sense, the biblical text has undoubtedly influenced the country's scientists and agronomists to question the intensive agriculture favoured in the early years of the State and to give high priority to conservation of land resources.

3.2 Cutting down fruit trees

When you are at war, and lay siege to a city . . . do not destroy its trees by taking the axe to them, for they provide you with food . . . (Deuteronomy 20:19).

In its biblical context this is a counsel of prudence rather than a principle of conservation; the Israelites are enjoined to use only 'non-productive', that is, non-fruit-bearing trees, for their siege works.

In rabbinic teaching, however, the verse has become the *locus classicus* for conserving all that has been created, so that the very phrase *bal tashchit*[36] is inculcated into small children to teach them not to destroy or waste even those things they do not need. In an account of the commandments specially written for his son, Rabbi Aaron Halevi of Barcelona (*c.* 1300) sums up the purpose of this one as follows:

This is meant to ingrain in us the love of that which is good and beneficial and to cleave to it; by this means good will imbue our souls and we will keep far from everything evil or destructive. This is the way of the devout and those of good deeds—they love peace, rejoice in that which benefits people and brings them to Torah; they never destroy even a grain of mustard, and are upset at any destruction they see. If only they can save anything from being spoilt they spare no effort to do so.[37]

3.3 *Limitation of grazing rights*

The Mishnah rules: 'One may not raise small cattle[38] in the Land of Israel, but one may do so in Syria or in the uninhabited parts of the Land of Israel'.[39] The history of this law has been researched[40] and there is evidence of similar restrictions from as early as the third century BCE.

The Mishnah itself does not provide a rationale for the law. Later rabbis suggest (a) that its primary purpose is to prevent the 'robbery' of crops by roaming animals and (b) that its objective is to encourage settlement in the Land. This latter reason is based on the premise that the raising of sheep and goats is inimical to the cultivation of crops, and reflects the ancient rivalry between nomad and farmer; at the same time it poses the question considered by modern ecologists of whether animal husbandry is an efficient way of producing food.[41]

3.4 *Agricultural festivals*

The concept of 'promised land' is an assertion that the consummation of social and national life depends on harmony with the land.

The biblical pilgrim-festivals all celebrate the Land and its crops, though they are also given historical and spiritual meanings. Through the joyful collective experience of these festivals the people learned to cherish the Land and their relationship, through God's commandments, with it; the sense of joy was heightened through fulfilment of the divine commandments to share the bounty of the land with 'the Levite, the stranger, the orphan and the widow' (Deuteronomy 16:11 and elsewhere).

4 SPECIFIC ENVIRONMENTAL LAWS

Several aspects of environmental pollution are dealt with in traditional halakha. Although the classical sources were composed in situations very different from the present the law has been, and is, in a continuous state of development, and in any case the basic principles are clearly relevant to contemporary situations.

4.1 *Waste disposal*

Arising from Deuteronomy 23:13, 14 halakha insists that refuse be removed 'outside the camp', that is, collected in a location where it will not reduce the quality of life. The Talmud and Codes extend this concept to the general prohibition of dumping refuse or garbage where it may interfere with the environment or with crops.

It would be anachronistic to seek in the earlier sources the concept of waste disposal as threatening the total balance of nature or the climate. However, if the rabbis forbade the growing of kitchen gardens and orchards around Jerusalem on the grounds that the manuring would degrade the local environment[42] one need have no doubt that they would have been deeply concerned at the large-scale environmental degradation caused by traditional mining operations, the burning of fossil fuels and the like. I would like to think that their response, had they been faced by the problem of disposal of nuclear wastes, would have led them to weigh up the evidence very carefully rather than to rush into an emotional judgement.

Smell (see also section 4.2) is regarded in halakha as a particular nuisance, hence there are rules regarding the siting not only of lavatories but also of odoriferous commercial operations such as tanneries.[43] Certainly, rabbinic law accords priority to environmental over purely commercial considerations.

4.2 *Atmospheric pollution and smoke*

Like smell, atmospheric pollution and smoke are placed by the rabbis within the category of indirect damage, since their effects are produced at a distance. They are nevertheless unequivocally forbidden.

The Mishnah[44] bans the siting of a threshing floor within 50 cubits of a residential area, since the flying particles set in motion by the threshing process would diminish the quality of the air.

Likewise, the second-century rabbi Nathan[45] rules that a furnace might not be sited within 50 cubits of a residential area because of the effect of its smoke on the atmosphere; the 50-cubit limit was subsequently extended by the Gaonim to whatever the distance from which smoke might cause eye irritation or general annoyance.[46]

The Hazards Prevention Law, passed by the Israeli Knesset on 23 March 1961, contains the following provisions:

> (3)   No person shall create a strong or unreasonable smell, of whatever origin, if it disturbs or is likely to disturb a person nearby or passerby.
>
> (4a) No person shall create strong or unreasonable pollution of the air, of whatever origin, if it disturbs or is likely to disturb a person nearby or passerby.

The subjectivity of 'reasonable' in this context is apparent. Meir Sichel, in a recent study[47] on the ecological problems that arise from the use of energy resources for power stations to manufacture electricity, and from various types of industrial and domestic consumption such as cooking, heating and lighting, has drawn on the resources of traditional Jewish law in an attempt to define more precisely what should be regarded as 'reasonable'. Citing rabbinic Responsa from an 800-year period he concludes that halakha is even more insistent on individual rights than the civil law (of Israel), and that halakha does not recognize 'prior rights' of a defendant who claims that he had established a right to produce the annoyance or pollutant before the plaintiff appeared on the scene.

It seems to me that in an exercise such as Sichel's there is no difficulty in applying traditional law to the contemporary context with regard to priority of rights, and also in clarifying the relationship between public and private rights. However, it is less clear that one can achieve a satisfactory definition of 'reasonable', since ideas of what is acceptable vary not only from person to person but in accordance with changing scientific understanding of the nature of the damage caused by smells and smoke, including the 'invisible' hazards of germs and radiation unknown to earlier generations.

4.3 Water pollution

Several laws were instituted by the rabbis to safeguard the freedom from pollution (as well as the fair distribution) of water. A typical early source:

> If one is digging out caves for the public he may wash his hands, face and feet; but if his feet are dirty with mud or excrement it is forbidden. [If he is digging] a well or a ditch [for drinking water], then [whether his feet are clean or dirty] he may not wash them.[48]

Pregnant with possibilities for application to contemporary life is the principle that one may claim damages or obtain an appropriate injunction to remove the nuisance where the purity of one's water supply is endangered by a neighbour's drainage or similar works. It is significant that the Gaonim here also rejected the talmudic distance limit in favour of a broad interpretation of the law to cover damage irrespective of distance.[49]

4.4 Noise

Rabbinic law on noise pollution offers a fascinating instance of balance of priorities. The Mishnah lays down that in a residential area neighbours have the right to object to the opening of a shop or similar enterprise on the grounds that the noise would disturb their tranquillity. It is permitted, however, to open a school for Torah notwithstanding the noise of children, for education has priority. Later authorities discuss the limit of noise which has to be tolerated in the interest of education,[50] and whether other forms of religious activity might have similar priority to the opening of a school.[51]

4.5 Beauty

Much could be said of the rabbinic appreciation of beauty in general. Here we concern ourselves only with legislation explicitly intended to enhance the environment, and we discover that it is rooted in the biblical law of the Levitical cities:

> Tell the Israelites to set aside towns in their patrimony as homes for the Levites, and give them also the common land surrounding the towns. They shall live in the towns, and keep their beasts, their herds, and all their livestock on the common land. The land of the

towns which you give the Levites shall extend from the centre of the town outwards for a thousand cubits in each direction. Starting from the town the eastern boundary shall measure two thousand cubits, the southern two thousand, the western two thousand, and the northern two thousand, with the town in the centre. They shall have this as the common land adjoining their towns (Numbers 35:2–5).

As this passage is understood by the rabbis, there was to be a double surround to each town, first a 'green belt' of 1,000 cubits, then a 2,000-cubit wide-belt for 'fields and vineyards'. Whilst some maintained that the 1,000-cubit band was for pasture, Rashi[52] explains that it was not for use, but 'for the beauty of the town, to give it space'—a concept reflected in Maimonides' interpretation of the talmudic rules on the distancing of trees from residences.[53] The rabbis debate whether this form of 'town planning' ought to be extended to non-Levitical towns, at least in the land of Israel, designated by Jeremiah (3:19) and Ezekiel (20:6, 15) 'the beautiful land'.

The rabbinic appreciation of beauty in nature is highlighted in the blessing they set to be recited when one sees 'the first blossoms in spring':

> You are blessed, Lord our God and ruler of the universe, who have omitted nothing from your world, but created within it good creatures and good and beautiful trees in which people may take delight.[54]

5 SAMPLE ETHICAL PROBLEMS RELATING TO CONSERVATION

5.1 Animal versus human life

Judaism consistently values human life more than animal life. One should not risk one's life to save an animal; for instance, if one is driving a car and a dog runs into the road it would be wrong to swerve, endangering one's own or someone else's life, to save the dog.

But is it right to take a human life, e.g. that of a poacher, to save not an individual animal but an endangered species? I can find nothing in Jewish sources to support killing poachers in any circumstances other than those in which they directly threaten human life. If it be argued that the extinction of a species would threaten human life because it would upset the balance of nature it is still unlikely that Jewish law would countenance homicide to avoid an indirect and uncertain threat of this nature.

Even if homicide were justified in such circumstances, how many human lives is a single species worth? How far down the evolutionary scale would such a principle be applied? After all, the argument about upsetting the balance of nature applies equally with microscopic species as with large cuddly-looking vertebrates like the panda, and with plants as much as with animals.

Judaism, true to the hierarchical principle of creation (section 2.4), consistently values human life more than that of other living things, but at the same time stresses the special responsibility of human beings to 'work on and look after' the created order (Genesis 2:15—see section 2.3).

5.2 Procreation versus population control

The question of birth control (including abortion) in Judaism is too complex to deal with here, but there is universal agreement that at least some forms of birth control are permissible where a potential mother's life is in danger and that abortion is not only permissible but mandatory up to full term to save the mother's life.[55] Significant is the value system which insists that, even though contraception may be morally questionable, it is preferable to abstinence where life danger would be involved through normal sexual relations within a marriage.[56]

What happens where economic considerations rather than life danger come into play? Here we must distinguish between (a) personal economic difficulties and (b) circumstances of 'famine in the world', where economic hardship is general.

On the whole, halakha places the basic duty of procreation above personal economic hardship. But what about general economic hardship, which can arise (a) through local or temporary famine and (b) through the upward pressure of population on finite world resources?

The former situation was in the mind of the third-century Palestinian sage Resh Lakish when he ruled: 'It is forbidden for a man to engage in sexual intercourse in years of famine'.[57] Although the ruling of Resh Lakish was adopted by the Codes[58] its application was restricted to those who already have children, and the decision between abstinence and contraception is less clear here than where there is a direct hazard to life.

Upward pressure of population on world resources is a concept unknown to the classical sources of the Jewish religion, and not indeed clearly understood by anyone before Malthus. As Feldman remarks:

> It must be repeated here that the 'population explosion' has nothing to do with the Responsa, and vice versa. The rabbis were issuing their analyses and their replies to a specific couple with a specific query. These couples were never in a situation where they might aggravate a world problem; on the contrary, the Jewish community was very often in a position of seeking to replenish its depleted ranks after pogrom or exile . . .[59]

Feldman goes on to say: 'It would be just as reckless to overbreed as to refrain from procreation'. Although I am not aware of any explicit traditional rabbinic source for this, I certainly know of none to the contrary. Indeed, as the duty of procreation is expressed in Genesis in the words 'be fruitful and multiply and fill the earth' it is not unreasonable to suggest that 'fill' be taken as 'reach the maximum population sustainable at an acceptable standard of living but do not exceed it'. In like manner the rabbis[60] utilize Isaiah's (45:18) phrase 'God made the earth . . . no empty void, but made it for a place to dwell in' to define the minimum requirement for procreation—a requirement, namely one son and one daughter, which does not increase population.

Of course, there is room for local variation amongst populations. Although as a general rule governments nowadays should discourage population growth there are instances of thinly populated areas or of small ethnic groups whose survival is threatened where some population growth might be acceptable even from the global perspective.

5.3 *Nuclear, fossil fuel, solar energy*

Can religious sources offer guidance on the choice between nuclear and fossil, and other energy sources?

It seems to me that they can have very little to say and that—especially in view of the extravagant views expressed by some religious leaders—it is vitally important to understand why their potential contribution to current debate is so small.

The choice among energy sources rests on the following parameters:

1 Cost effectiveness.
2 Environmental damage caused by production.
3 Operational hazards.
4 Clean disposal of waste products.
5 Long-term environmental sustainability.

Let us consider these parameters. Cost effectiveness cannot be established without weighing the other factors. There is no point, however, at which religious considerations apply in establishing whether a particular combination of nuclear reactor plus safety plus storage of waste and so on will cost more or less than alternative 'packages' for energy production.

It is equally clear that religious considerations have no part to play in assessing environmental damage caused by production, operational hazards, whether waste products can be cleanly disposed of, or what is the long-term environmental sustainability of a method of energy production. These are all technical matters, demanding painstaking research and hard evidence, and they have nothing to do with theology.

The religious might perhaps have something to say about overall strategy. For instance, a religious viewpoint might suggest that scientists should pay more attention to finding out how to use less energy to meet demands for goods than to finding out how to produce more energy. However, unless the religious are actually aiming to persuade people to demand fewer goods, such advice—viz. to seek more energy-efficient ways to do things—is merely the counsel of prudence, not dependent on any characteristically religious value.

It is a matter of sadness and regret that religious leaders are so prone to stirring up the emotions of the faithful for or against some project, such as nuclear energy, which really ought to be assessed on objective grounds. Much of the hurt arises from the way the religious 'demonize' those of whom they disapprove, and in the name of love generate hatred against people who seek to bring benefit to humanity.

5.4 Global warming

A very similar analysis could be made of the problems relating to global warming—problems of which scientists have been aware since Arrhenius in the late nineteenth century, though only recently have pressure groups developed and governments become alarmed. The fact is that in 1992 no one knows the extent, if any, to which global temperatures have risen as a result of the rise in atmospheric carbon dioxide from 290 parts per million in 1880 to 352 parts per million in 1989, and no one knows what would be the overall effects of the projected doubling of atmospheric carbon dioxide by the middle of next century (I leave aside the question of other greenhouse gases). Some consequences, indeed, may be beneficial, such as greater productivity of plants in an atmosphere with more carbon dioxide. Unfortunately, neither the techniques of mathematical modelling used to make the projections, nor the base of global observations at 500 kilometre intervals, can yield firm results.[61] So how can a government decide whether to spend hundreds of billions of dollars on reducing atmospheric carbon dioxide, and vast sums in aiding Third World countries to avoid developing along 'greenhouse' lines, when the draconian measures required greatly limit personal freedom and much of the expenditure might be better diverted to building hospitals, improving education and the like? Essential steps, including better research, must be initiated, but it would be a lack of wisdom to rush into the most extreme measures demanded. From our point of view, however, it is clear that the decisions must be rooted in prudence, rather than in any specifically religious value (of course, all religions commend prudence).

5.5 *Who pays the piper?*

Our observations on the response to the possibility of global warming raised the question of paying for conservation. The dilemmas involved in this are exceedingly complex. Should rich nations pay to 'clean up' the technology of poorer nations (e.g. Western Europe pay for Eastern Europe)? Should governments distort the free market by subsidizing lead-free petrol and other 'environment-friendly' commodities? How does one assess environmental efficiency and social costs, and how should such costs be allocated between taxpayer, customer and manufacturer?

5.6 *Directed evolution*

After writing about the progress from physical evolution through biological evolution to cultural evolution, Edward Rubinstein continues:

> Henceforth, life no longer evolves solely through chance mutation. Humankind has begun to modify evolution, to bring about non-random, deliberate changes in DNA that alter living assemblies and create assemblies that did not exist before.
>
> The messengers of directed evolution are human beings. Their messages, expressed in the language and methods of molecular biology, genetics and medicine and in moral precepts, express their awareness of human imperfections and reflect the values and aspirations of their species.[62]

These words indicate the area where religions, Judaism included, are most in need of adjusting themselves to contemporary reality—the area in which modern knowledge sets us most apart from those who formed our religious traditions. Religion as we know it has come into being only since the Neolithic revolution, and thus presupposes some technology,[63] some mastery of nature. But it has also assumed that the broad situation of humanity is static, and this is now seen to be an illusion.

All at once there is the prospect, alarming to some yet challenging to others, that we can set the direction of future development for all creatures in our world. The Ethics Committees of our hospitals and medical schools are forced to take decisions;

43

although the religious take part—and Judaism has a distinctive contribution to make to medical ethics[64]—it has yet to be shown that traditional sources can be brought to bear other than in the vaguest way ('we uphold the sanctity of life') on the problems raised even by currently available genetic engineering.

Will religions, as so often in the past, obstruct the development of science? They need not. Jewish religious views have ranged from Isaac Abravanel, who opposed in principle the development of technology,[65] to Abraham bar Hiyya, who in the twelfth century played a major role in the transmission of Graeco–Arab science to the West. If Judaism (or any other religion) is to contribute towards conservation it will need to be in the spirit of bar Hiyya, through support for good science, rather than through idealization of the 'simple life' in the spirit of Seneca and Abravanel.

6 CONCLUSION—RELIGION AND CONSERVATION

There is no doubt that Judaism, along with other religions, has resources which can be used to encourage people in the proper management of Planet Earth. We will now review the interaction of religion with conservation with special reference to the sources cited.

1 We saw in section 2.1 how Judaism interprets the created world, with its balanced biodiverse ecology, as a 'testimony to God', with humankind at the pinnacle holding special responsibility for its maintenance and preservation. Certainly, this attitude is more conducive to an interest in conservation than would be emphasis on the centrality of the 'next world', on the spirit versus the body, or on the 'inferior' or 'illusionary' nature of the material world.

2 One of the priorities of conservation at the present time is to control population so as not to exceed resources. Although Judaism stresses the duty of procreation we learned in section 5.2 that it offers the prospect of constructive approach to popu-

lation planning, including some role for both contraception and abortion.

3 We have noted several specific areas in which Judaism has developed laws or policies significant for conservation. Prime among them (section 3) were the laws regulating the relationship between people and land, for which the 'chosen people' in the 'promised land' is the model. Care of animals (section 2.5), waste disposal, atmospheric and water pollution, noise, and beauty of the environment (section 4) were also treated in the classical sources. It would be neither possible nor fully adequate to take legislation straight from these sources; but it is certainly possible to work in continuity with them, bearing in mind the radically new awareness of the need for conserving the world and its resources as a whole.

4 Religions, Judaism included, discourage the pursuit of personal wealth. Whilst in some instances this may be beneficial to the environment—if people want less cars and less books there will be less harmful emissions and less forests will be chopped down—there are also many ways in which poverty harms the environment—for instance, less research and development means that such technology as remains (presumably for hospitals and other welfare projects) will be less efficient and the problems of environmental pollution less effectively addressed. It is a moot point whether lower technologies generate less pollution *pro rata* than advanced ones.

5 Some religions remain strongly committed to evangelistic or conversionist aims which inhibit co-operation with people of other religions. Judaism is not currently in an actively missionary phase; some would say that it is unduly introspective, and needs to proclaim its values in a more universal context. All religions, however, must desist from ideological conflicts and espouse dialogue; conservation cannot be effective without global co-operation.

6 Mere information can motivate, as when someone who perceives a lion ready to pounce reacts swiftly. If ecological disaster were as clearly perceived as a crouching lion ideological

motivation would be unnecessary. It is better that religions support conservation than oppose it, but the world would be safer if people would act on the basis of rational collective self-preservation rather than on the basis of confused and uncontrollable ideologies.

7 Several times, particularly in discussing energy sources in section 5.3 and global warming in section 5.4, we had to stress the need to distinguish between technological and value judgements. Whether or not nuclear reactors should be built must depend on a careful, dispassionate assessment of their hazards; shrill condemnation of the 'hubris of modern technology' merely hinders judgement, though it is right and proper that religious values be considered when an informed choice is made.

Of course, the same need for objective assessment before value judgements are made applies to all other major conservation questions, such as how to reverse deforestation, control the greenhouse effect, restore the ozone layer.

8 Towards the end of section 5.6 we noted a characteristic religious ambivalence towards science. In the interests of conservation it is essential that the 'pro-science' attitude of Abraham bar Hiyya, Maimonides and others be encouraged. The craziness of 'simple life' proponents must be resisted. For a start, the present world population could not be supported if we were to revert to the simple life. Moreover, who would wish to do without sanitation, communications, electric light, books, travel, medical services and all those other benefits of 'complex' civilization? The small population which would survive the 'return to Eden' would live a very dull and insecure life.

There is indeed a question as to what level of technology—simple, intermediate, advanced—is most appropriate in a given situation. We should not be surprised if scientific evidence indicates that in some situations simple technology is preferable to advanced, whether because of availability of skills and resources, or whether because of side effects such as pollution. Once again, the moral issue is straightforward (one should achieve the most satisfactory balance between people's wants and the conservation of nature), but the decision has to be based on sound technical and scientific evidence.

If science has got us into a mess (which I would dispute) the way out is not *no* science but *better* science, and science performed with a sense of moral responsibility.

Finally, let us note that Judaism, like other religions, has a vital role to play in eradicating those evils and promoting those values in society without which no conservation policies can be effective. The single greatest evil is official corruption, frequently rife in precisely those countries where conservation measures must be carried out. Next in line is drug addiction with its associated trade. Religions must combat these evils and at the same time work intelligently for peace, not only between nations but amongst religions themselves.

USEFUL QUOTATIONS AND ANECDOTES

FROM THE BIBLE

Translations are mainly those of the New Revised Standard Version, with some modifications.

God created humans in his image . . . male and female he created them. (Genesis 1:27)

God saw everything he had made, and indeed it was very good. (Genesis 1:31)

The Lord God took the man and put him in the garden of Eden to till it and to preserve it (Genesis 2:15). The man gave names to all the cattle, and to the birds of the air, and to every animal of the field. (Genesis 2:20, meaning that the man understood the nature of each of the animals.)

When you are at war, and lay siege to a city . . . do not destroy its trees by taking the axe to them, for they provide you with food . . . (Deuteronomy 20:19—the basis of *bal tashchit*, 'do not waste'.)

47

With your utensils you shall have a trowel; when you relieve yourself outside, you shall dig a hole with it and then cover up your excrement. Because the Lord your God travels along with your camp . . . (Deuteronomy 23:13–14, on which halakha bases the insistence that refuse be removed 'outside the camp', that is, collected in a location where it will not reduce the quality of life.)

> The earth is the Lord's and all that is in it,
>> the world, and those who live in it,
> for he has founded on the seas,
>> and established it on the rivers. (Psalm 24:1)

> You make springs gush forth in the valleys;
>> they flow between the hills,
> giving drink to every wild animal;
>> the wild asses quench their thirst.
> By the streams the birds of the air have their habitation;
>> they sing among the branches.
> From your lofty abode you water the mountains;
>> the earth is satisfied with the fruit of your work.
> You cause the grass to grow fat for the cattle,
>> and plants for people to use. (Psalm 104:10–14)

FROM THE RABBIS

You are blessed, Lord our God and Ruler of the universe, who have omitted nothing from your world, but created within it good creatures and good and beautiful trees in which people may take delight. (Prayer Book)

A calf was about to be slaughtered. It ran to Rabbi (Judah the Patriarch, about the year 200) and nestled its head in his robe and whimpered. He said to it, 'Go! This is what you were created for!' As he had no mercy on it, heaven decreed suffering upon him. One day Rabbi's housekeeper was sweeping. She came across some young weasels and threw them and swept them out; he said, 'Let them alone! Is it not written, "His mercies extend to all his creatures" (Psalm 148:9)?' Heaven decreed, 'Since he is merciful, let us show him mercy'. (Slightly adapted from B. *Bava Metzia* 85a)

A comment of the twelfth-century Jewish Bible commentator Abraham ibn Ezra on the words of Psalm 115:16, 'The heavens are the heavens of the Lord, and he gave the earth to people': 'Ignorant people have compared man's rule over the earth with God's rule over the heavens. This is not right, for God rules over everything. The meaning of "he gave it to people" is that man is God's steward (*paqid*—officer or official with special responsibility for a specific task) over the earth, and must do everything according to God's word.'

This is meant to ingrain in us the love of that which is good and beneficial and to cleave to it; by this means good will imbue our souls and we will keep far from everything evil or destructive. This is the way of the devout and those of good deeds—they love peace, rejoice in that which benefits people and brings them to Torah; they never destroy even a grain of mustard, and are upset at any destruction they see. If only they can save anything from being spoilt they spare no effort to do so. (Rabbi Aaron Halevi of Barcelona, *c.* 1300, explaining *bal tashchit*.)

Joseph Albo wrote that just as clothes are an integral part of the animal, but external to people, who have to make clothes for themselves, so are specific ethical impulses integral to the behaviour of particular animals, and we should learn from their behaviour. 'Who teaches us from the beasts of the earth, and imparts wisdom to us through the birds of the sky' (Job 35:11)—as the Talmud puts it: 'R. Yohanan said, If these things were not commanded in the Torah, we could learn modesty from the cat, the ant would preach against robbery, and the dove against incest'. The superiority of humans lies in their unique combination of freedom to choose and the intelligence to judge, without which the divine Revelation would have no application. Being in this sense 'higher' than other creatures, humans must be humble towards all. Albo, in citing these passages and commending the reading of *Pereq shira*, articulates the attitude of humble stewardship towards creation which characterizes rabbinic Judaism.

Rav Kook (1865–1935), drawing on a range of classical Jewish sources from Psalm 148 to Lurianic mysticism, beautifully acknowledges the divine significance of all things (the immanence of God): 'No rational person can doubt that the Torah, when it commands people to "rule over the fishes of the sea and the birds of the sky and all living things that move on the earth" does not have in mind a cruel ruler who exploits his people and servants for his own will and desires—God forbid that such a detestable law of slavery [be attributed to God] who "is good to all and his tender care rests upon all his creatures" (Psalm 145:9) and "the world is built on tender mercy" (Psalm 89:3)'. (Rav Kook on Genesis 1:28)

I recall that with God's grace in the year 5665 [1904/05] I visited Jaffa in the Holy Land, and went to pay my respects to its Chief Rabbi [Rav Kook]. He received me warmly ... and after the afternoon prayer I accompanied him as he went out into the fields, as was his wont, to concentrate his thoughts. As we were walking I plucked some flower or plant; he trembled, and quietly told me that he always took great care not to pluck, unless it were for some benefit, anything that could grow, for there was no plant below that did not have its guardian above. Everything that grew said something, every stone whispered some secret, all creation sang ... (A reminiscence by Aryeh Levine)

NOTES

1 Much of the material in this section first appeared as Norman Solomon, 'Judaism and conservation', *Christian Jewish Relations* XXII, Summer 1989. See also Norman Solomon, *Judaism and World Religion*, Macmillan, London/ St Martin's Press, New York, 1991, chapter 2.

2 Lichtenstein's paper was published (in Hebrew) in *Haggut* V, Religious Education Department of the Israel Ministry of Education, Jerusalem 5740 (1980), pp. 101–8.

3 See Samuel Cooper's anthropological study of the 'laws of mixture' in Harvey E. Goldberg (ed.), *Judaism: Viewed from Within and from Without*, State University of New York Press, Albany, 1987, chapter 1. Our approaches are radically different but not mutually exclusive.

4 See the special 'World Ships' issue of the *Journal of the British Interplanetary Society* XXXVII, June 1984, for some detailed engineering observations. Alexander G. Smith's article 'Worlds in miniature—Life in the starship' explores the ecological aspects of such Arks.

5 To the third-century Palestinian Rabbi Abbahu the Midrash *Bereshit Rabbah* 3:9 attributes the statement that God 'created and destroyed worlds before he made these'—which are presumably his final, perfect design.

6 Malachi Beit-Arieh prepared a critical edition of the work as a PhD thesis, Jerusalem 1966, but it has not been published. Our remarks centre on the texts in the large *Siddurim* (Prayer Books), particularly those of Jacob Emden and Seligmann Baer.

7 Seligmann Baer (1825–1907), in his introduction to *Pereq shira* on p. 547 of his masterly 1868 edition of the Prayer Book *Seder Avodat Israel*, correctly denies the authenticity of these introductory dicta, and omits them from his edition.

8 See David Ehrenfeld and Philip J. Bentley, 'Judaism and the practice of stewardship', *Judaism* XXXIV, 1985, pp. 310–11.

9 Abraham Isaac Kook was born in Latvia and emigrated to Palestine in 1904, becoming Chief Rabbi of the Ashkenazi communities of Palestine when the office was instituted in 1921. A man of great piety and erudition, his numerous works are imbued with mysticism, and he emphasized the role of holiness in establishing the Jewish presence in the Holy Land. A selection of his writings translated into English is published under the title *Abraham Isaac Kook* in the series The Classics of Western Spirituality: Paulist Press, New York, 1978.

10 This is a traditional Jewish understanding of the text. Versions such as 'For ever is mercy built' (translation of the Jewish Publication Society of America, consonant with several English versions) are grammatically sounder.

11 I have taken the quotation from the texts on *Protection of Animals* published (in Hebrew) by the Israel Ministry of Justice in February 1976, but have been unable to check the original source. This publication of the Ministry of Justice together with its volume on *Protection of the Environment* (July 1972) is an excellent resource for traditional texts on these subjects, having been compiled to assist those responsible for drafting legislation for the Knesset. The volumes were compiled by Dr Nahum Rakover, in his capacity as Adviser on Jewish Law to the Ministry of Justice.

12 In view of the ending of the verse this is a more appropriate translation of Hebrew *adam*, a generic term for humankind, than the sexist 'man'.

13 Joseph Albo, *Sefer Ha-Iqqarim*, Book III, chapter 1.

14 B. *Eruvin* 100b. We respect Yohanan's reverence for nature, not his skill in scientific observation.

15 Heb. *mazzal*—literally 'constellation', but understood also as 'guardian angel'.

16 A reminiscence by Aryeh Levine in *Lahai Roi* (in Hebrew), Jerusalem, 5721 (1961), pp. 15, 16.

17 The 'Seven Laws of the Children of Noah' attempt to define the religious obligations of humankind in general, for all people are descended from Noah. The laws, unknown in this form in sources earlier than the third century, are: Do not blaspheme, do not worship idols, do not murder, do not commit adultery, do not steal, do establish courts of justice, do not eat 'a limb torn from a living animal'. The last of these covers cruelty to animals; it is well explained in chapter 8 of David Novak, *The Image of the Non-Jew in Judaism* (Toronto Studies in Theology, no. 14), Edwin Mellen Press, New York and Toronto, 1983.

18 B. *Bava Metzia* 85a and *Genesis Rabbah* 33.

19 Moses E. Gaster (ed.), *Masseh Book*, Vol. I, Jewish Publication Society of America, Philadelphia, 1934.

20 In B. *Shabbat* 128b it is suggested that this principle is of biblical status (*d'oraita*).

21 B. *Berakhot* 40a. See *Orach Hayyim* 167:6.

22 See Elijah Judah Shochet, *Animal Life in Jewish Tradition: Attitudes and Relationships*, New York, 1984.

23 See J. David Bleich, *Contemporary Halakhic Problems*, Vol. 3, KTAV, New York, 1989, pp. 194–236, for a review of the halakhic literature on animal experimentation.

24 Joseph Albo, *Sefer Ha-Iqqarim* 3:15.

25 Isaac Abravanel, *Commentary on Isaiah* (in Hebrew), chapter 11 on the verse 'The wolf shall lie down with the lamb'.

26 Isaac Abravanel, *Commentary on Genesis* (in Hebrew), chapter 2.

27 See Richard Schwarz, *Judaism and Vegetarianism*, Exposition Press, 1982. See also J. David Bleich, *Contemporary Halakhic Problems*, Vol. 3, KTAV, New York, 1989, pp. 237–250b for a review of the halakhic literature on vegetarianism.

28 In his commentary *Baer Heitev* on *Shulkhan Arukh: Orach Hayyim*, chapter 134, note 3.

29 Known as the 'Ponevezher Rav', from the Lithuanian town where he established his reputation as a yeshiva lecturer, Kahaneman survived the Holocaust and spent his latter years in Israel where he built up a network of yeshivot, orphanages and other institutions. My information on his diet was received in conversation with his disciples.

30 The reference to his text on the Protection of Animals is above, and the relevant section commences on p. 7. He gives a wide range of references to the Responsa, many of which come from Renaissance Italy which provided most of the very few instances of Jews in pre-modern times engaging in hunting.

31 Since the prey, even if a kosher animal, will not have been slaughtered lawfully.

32 Time should be devoted to study and good deeds.

33 He implies that participating in hunting takes one out of the company of Torah scholars and into that of those who mock at religious values.

34 The point is that hunting is contrary to the Jewish ethos.

35 For an interesting account of tension within the Islamic world between religious teaching and social norms on the subject of hunting see John R. Willis, 'Hunting, hawking and learning: "manly" attainments and the pursuit of knowledge in early Islam', *Maghreb Review* XIII, 1988.

36 Literally, 'not to destroy'. In B. *Makkot* 22a Ravina (fourth century) stresses the positive aspect of the commandment, 'but you *shall* eat (the fruit of the trees)'.

37 *Sefer Ha-Hinnukh* Mitzvah 529.

38 Sheep, goats etc.

39 M. *Bava Qama* 7:7.

40 See for instance G. Allon, *'Toldot ha-Yehudim b'Eretz Israel bi-Tequfat ha-Mishnah v'ha-Talmud*, Hakibbutz Hameuchad, Tel Aviv, 1953/5, Vol. I, pp. 173–8 and 359.

41 Some caution is needed here. The rabbis of the Talmud did not envisage vegetarianism, and did not ban the raising of large cattle in the Land. They assumed that meat would be eaten but tried to ensure that its production would not interfere with agriculture.

42 B. *Bava Qama* 82b.

43 These matters are dealt with in the Talmud in the second chapter of *Bava Batra*. They are codified, with subsequent developments, in *Shulkhan Arukh: Hoshen Mishpat*, chapter 145. Maimonides, in his philosophical work *Guide For The Perplexed*, Book 3, chapter 45, argues that the purpose of the incense

in the Temple was to counteract the smell of the processing of the animal offerings.

44 M. *Bava Batra*, chapter 2.

45 T. *Bava Batra* 1:7.

46 A. Assaf (ed.), *Teshuvot ha-Gaonim'*, Darom, Jerusalem, 5689 (1929), p. 32. The Gaonim were the heads of the Babylonian academies after the completion of the Talmud; they occupy a major place in the development and transmission of rabbinic law.

47 Meir Sichel, 'Air pollution—Smoke and odour damage', *Jewish Law Annual*, V, Boston University and E.J. Brill, Leiden, 1985, pp. 25–43. We have used Sichel's translation of the Israeli legislation referred to.

48 T. *Bava Metzia* 11:31 (ed. Zuckermandel).

49 Cited in *Shulkhan Arukh: Hoshen Mishpat* 155:21.

50 Rashi on B. *Bava Batra* 21a. Nahmanides, in his commentary on the passage, hazards a guess that the permissible noise limit would be exceeded by a school of more than 50 pupils.

51 *Shulkhan Arukh: Hoshen Mishpat* 156:3.

52 On B. *Sota* 22b.

53 See Maimonides, *Mishneh Torah: Shekhenim*, chapter 10.

54. In the name of Judah bar Ezekiel (third-century Palestinian) in B. *Berakhot* 43b. A whole chapter of *Shulchan Arukh—Orach Hayyim* 226—is devoted to it.

55 For a full treatment of these issues see David M. Feldman, *Marital Relations, Birth Control and Abortion in Jewish Law*, Schocken Books, New York, 1974.

56 Feldman, ibid., p. 302.

57 B. *Taanit* 11a.

58 *Shulkhan Arukh: Orach Hayyim* 240:12 and 574:4.

59 Feldman, op. cit., p. 304.

60 B. *Yevamot* 62a.

61 In this section we are indebted to the summary provided by Robert M. White, 'The great climate debate', *Scientific American*, July 1990.

62 E. Rubinstein, 'Stages of evolution and their messengers', *Scientific American*, June 1989, p. 104.

63 The ancients thought this had come from the gods, but Genesis 4:20–22 polemically, if more accurately, credits humans with technological innovation.

64 See, in addition to David Feldman, op. cit., I. Jakobovits, *Jewish Medical Ethics*, 4th ed., New York, 1975. Huge numbers of Responsa appear in halakhic journals, and there are specialist periodicals on Jewish medical ethics in Hebrew and English. The Federation of Jewish Philanthropies of New York issues frequent updates (for instance, 6th ed. 1984) of its *Compendium on Medical Ethics*, which reflects a broad consensus of views across the Jewish denominations.

65 See his commentary on Genesis 2. He taught that in the Messiah's time, as in Eden, we would wear no clothes, build no houses, abandon technology and have no government; in this he is more indebted to Seneca's 90th Epistle than to Jewish sources.

4 TRADITIONAL JEWISH ATTITUDES TOWARDS PLANT AND ANIMAL CONSERVATION

Yosef Orr and Yossi Spanier

It is remarkable how ancient sources govern modern life. Israel's generals, in time of war, have frequently used biblical information as to ancient paths and topographical features in deciding on strategy and tactics. Similarly the authors of the article that follows hark back constantly to the old teachings as an aid to modern assessment of Israel's flora and fauna. They emphasize the interdependence of all forms of life, a fact increasingly understood as we survey the consequences of an artificially created imbalance in the living world.

The writers have been Field Officers of the Society for the Protection of Nature in Israel (SPNI), an organization devoted to research and protection of all aspects of nature in that country. The organization has not been slow in resisting assaults on the environment. They are part of a discipline known in Israel as yediat ha–aretz *(knowledge of the country), Israelography, now a recognized field of study at all levels in the educational system.*

Contact with the Society is a must for any botanist, geologist, zoologist, or nature-lover, or anyone concerned with the environment.

In the ongoing effort to preserve the plants and animals of the land of Israel, sayings of the Jewish sages may serve as a constant source of encouragement and support. Jewish tradition and folklore, going back as far as the Torah—the Pentateuch—repeatedly refer to the integral links between all aspects of the natural world, viewing man as an intrinsic part of the entire fabric.

Any examination of the traditional Jewish attitude towards nature conservation must begin with an identification of the values that are considered important and worthy of protection. The land of Israel itself has pride of place. The Bible describes the country with its natural riches:

> For the Lord thy God bringeth thee into a good land, a land of brooks of water, of fountains and depths, springing forth in valleys and hills; a land of wheat and barley, and vines and fig-trees and pomegranates; a land of olive trees and honey; a land wherein thou shalt eat bread without scarceness, thou shalt not lack any thing in it; a land whose stones are iron, and out of whose hills thou mayest dig brass. (Deuteronomy 8:7–9)

> A land which the Lord thy God careth for; the eyes of the Lord thy God are always upon it, from the beginning of the year even unto the end of the year. (Deuteronomy 11:12)

The climatic variety of the land, and the striking transitions between landscapes of seashores and deserts are echoed in the words of the sages:

> The land of Israel ... has everything; for all other countries each have this feature of the world, or a different one; but the land of Israel lacks nothing, as is stated in Deuteronomy (8:9): 'Thou shalt lack nothing in it'. (Sifri, *Eqev* 37)

The land is famous for its beauty as well: 'The nature of the land of Israel is to be more beautiful than all other lands' (Yerushalmi, *Ketubot* 13:6).

The sages appreciated the pleasure that nature provides to all the senses: 'Three things may strengthen a person: sound, sight, and smell' (*Berakhot* 57b). 'What brings pleasure to the soul rather than to the body? Smell' (*Berakhot* 43b).

Since nature brings pleasure, the sages were concerned over every injury to it; anyone who spoiled it was transgressing an explicit prohibition: 'Anyone who spoils something that gives pleasure is transgressing against the injunction "thou shalt not spoil"' (*Midrash Aggada*, Judges 20).

PLANT CONSERVATION

The dramatic processes of plant growth and leaf-budding in spring are so impressive that the sages formulated a special blessing to mark one's first encounter with trees in blossom:

> Rabbi Yehuda said: He who goes out in the month of Nissan and sees trees coming into leaf says, Blessed be He who has made the world so that it lacks nothing, and created good creatures and good trees, for the pleasure of men. (*Berakhot* 43b)

In a different blessing, trees are included among living creatures as equals: 'He who sees goodly creatures and fine trees says, Blessed be He who has such in His world' (*Berakhot* 58a).

The comparison goes beyond general terms. Trees, particularly those which bear fruit, are likened to humans: For man is as the tree of the field: this indicates that man's life depends on the trees ... He who uproots a tree is transgressing against the prohibitions' (Sifri, Judges 23).

People naturally tend to be more careful of fruit trees; but the sages enjoined us to take care of non-fruit-bearing trees as well, since they mark farm boundaries and thus prevent intrusions into personal domain: 'He who wants his property to flourish plants an "eder" [a type of non-fruit-bearing tree], since a field with an "eder" in it is not robbed' (*Betza* 15b).

The sages encouraged people to grow plants within the bounds of settlements and formulated an original precept: 'Rabbi Huna said, A scholar is forbidden to live in a town that has no plants' (*Eruvim* 55b), because 'a dwelling place should be beautiful in the eyes of its inhabitants' (*Sota* 47a). Since the residents of any town were interested in attracting scholars, they would supposedly lead to the planting of more trees.

The view of trees as living creatures, and their comparison with man, gave rise to severe injunctions against harming trees. The seriousness of chopping down trees is expressed picturesquely: 'When a fruit bearing tree is chopped down, a voice is heard from one end of the world to the other but it is not audible' [to the human ear] (*Pirkei deRabi Eliezer* 34). Not only is the voice of a wounded tree heard over a great distance; the people who caused the injury will be punished: 'Those who chop beneficial trees ... will never be blessed in their work' (*Pesahim* 50b).

In many cultures, an eclipse of the sun or of the moon bodes ill. One Jewish tradition considers them a punishment for unjust acts. One reason for eclipses is the destruction of trees: 'The lights of the world suffer ... because of the destroyers of beneficial trees' (*Sukkah* 29a).

The righteous man is likened to a tree in order to illustrate the gravity of uprooting a fruit tree: 'Just as an uprooted date palm cannot be replaced, a righteous person cannot be replaced' (*Numbers Rabba* 3a).

PROTECTING FRUIT TREES

Even in wartime, when many prohibitions seem to be lifted or eased, the Torah forbids the destruction of fruit trees:

> When thou shalt besiege a city a long time, in making war against it, thou shalt not destroy the trees thereof by wielding an axe against them; for thou mayest eat of them, but thou shalt not cut them down; for is the tree of the field man, that it should be besieged of thee? Only the trees of which thou knowest that they are not trees for food, them thou mayest destroy and cut down, that thou mayest build bulwarks against the city that maketh war with thee, until it fall. (Deuteronomy 20:19–20)

The sages set up a law against grazing by goats, in order to protect the natural forest of the country. The comparison of tree-destroyers with grazing goats is clear in the following: 'The light sources of the world are eclipsed for four reasons ... and because of goat-raisers in the Land of Israel' (*Sukkah* 29a). The sages were apparently fully aware of the damage caused by uncontrolled grazing by goats, and decreed: 'Goats may not be raised in the land of Israel; they may be raised in Syria and in the deserts of the Land of Israel' (*Bava Kama* 79b).

This view is strengthened by the mention of King David, who was a shepherd in the desert before he became king: '... with whom has thou left those few sheep in the wilderness?' (1 Samuel 17:28). David is considered to have followed the law and raised his flocks only in the desert areas of the country (*Exodus Rabba* 2:3).

The Talmud tells the following story, to emphasize the negative attitude towards the 'destroyer of the forest':

> A pious man had a bad cough. The physicians said that the only cure was to drink fresh boiled milk. A goat was brought to the house (despite the prohibition) and tied to a leg of the bed, so that the sick man was able to drink fresh milk. When his friends came to visit, they saw the goat and turned away, saying 'The goat is like an armed robber in the house—why should we enter?' This was the only sin they could ascribe to the man, and he himself, on his deathbed, said 'I know that this was my only sin'. (*Bava Kama* 80a)

ATTITUDE TOWARDS ANIMALS

The dependence of man's existence on animals—'a man's domestic animals are his very life' (*Exodus Rabba* 26:2)—has always been clear. The custom of hunting fowl is known from the following legend:

> The vulture [eagle] is different from all other birds. All the other birds hide their young under their feet, for fear of another bird flying overhead. But the eagle is only afraid of man and the arrows he shoots. (Mekhilta, *Yitro* 29)

The Torah emphasizes man's responsibility for the continued existence of animal species:

> If a bird's nest chance to be before thee in the way, in any tree or on the ground, with young ones or eggs, and the dam sitting upon the young, or upon the eggs, thou shalt not take the dam with the young; thou shalt in any wise let the dam go, but the young thou mayest take unto thyself; that it may be well with thee, and that thou mayest prolong thy days. (Deuteronomy 22:6–7)

The first biblical mention of the threat of extinction—and of the preservation—of animals is in the story of Noah: 'And of every living thing of all flesh, two of every sort shalt thou bring into the ark, to keep them alive with thee; they shall be male and female' (Genesis 6:19). The sages emphasize that 'the Lord did not think of Noah alone; He remembered Noah and all the beasts and the animals with him' (*Pesikta Zutreta*, Genesis 8:1).

Noah, the most righteous man of his generation, provides an example of the proper treatment of animals: 'Throughout those twelve months, Noah and his sons did not sleep, because they had to feed the animals, the beasts and the birds' (*Tanhuma* 58:9). This behaviour is the basis for all instructions and guidelines educating towards consideration for animals: 'No man may take an animal, a beast or a bird, unless he provides them with food' (Yerushalmi, *Yevamot* 15:3); 'No man may eat before feeding his animals' (*Berakhot* 40a1); 'It is a good sign for a man when his animals eat and are satisfied' (Sifri, *Eqev* 43).

The duty of mercy applies to all animals, including those belonging to a non-Jew: 'The animal of an idolater should be looked after in the same way as the animals of Israelites' (*Bava Matzi'a* 32b).

The Torah concerns itself with the pain suffered by animals, and ordains that care should be taken to prevent it: 'Thou shalt not plough with an ox and an ass together' (Deuteronomy 22:10); 'Thou shalt not see thy brother's ass or his ox fallen down by the way, and hide thyself from them; thou shalt surely help him to lift them up again' (Deuteronomy 22:4).

Every creature has the right to exist.

> Rabbi Yehuda said, in the name of Rav, Everything the Lord created in His world has a purpose—even the things that man may consider to be unnecessary, such as flies, fleas and mosquitoes, are part of the creation. (*Genesis Rabba* 10:8)

The world exists thanks to living creatures—and for their sake: 'It is not thanks to you that rain falls, or that the sun shines—it is thanks to the animals' (*Genesis Rabba* 33:1).

In tradition, the Creator himself serves as an example of care and consideration for all created things: 'Just as the Lord feels pity for man, He has pity for animals' (*Devarim Raba* 6:1). 'The Most Righteous understands the soul of animals, even when He is angry' (*Tanhuma* 58:7).

The Creator's personal care for created things is clear in two examples pertaining to the natural world: 'When a female ibex is about to give birth, she climbs a hill. I [the Creator] call up an eagle to receive the new-born ibex on its wings, and the eagle is always there just in time' (*Bava Batra* 16a–b). The birth process of the ibex (*Capra ibex nubriana*) in fact takes place on clifftops, and the Creator's personal care for the newborn can be construed as a parable.

A similar case is the description of the birth of a gazelle; this should also be understood as awareness of the miracle of birth through a narrow passage:

> The body of the gazelle is slender, and she has trouble giving birth. What does the Creator do? He calls up a snake, which bites the gazelle enough to relax her entire body. What does the Creator do after the gazelle gives birth? He brings her a herb which she eats; this helps her to recover from the birth. (*Yalkut Shim'oni*, Psalms, 104)

Beyond the economic dependence of man on natural resources, the traditional Jewish view holds that the living world is a single entity, and that life is a value to be preserved in animals as well as in man:

> For that which befalleth the sons of men befalleth beasts; even one thing befalleth them; as the one dieth, so dieth the other; yea, they have all one breath; so that man hath no pre-eminence above a beast ... (Ecclesiastes 3:19)

All creatures are interdependent, and none can exist without the others. 'See, how all the Lord's creatures borrow from one another' (*Exodus Rabba* 31:15). The emphasis is on 'borrow', rather than 'take' ... The fabric of food in the natural world consists of creatures borrowing the materials to build their bodies out of natural components, and returning them, like any honest borrower, when the loan falls due.

The inescapable conclusion is that harm to any creature, as well as to nature as a whole, affects the entire fabric of creation. There is more than a hint here of an ecological philosophy. All living creatures face the same destiny.

> The whole world, man, animals and birds, all find their food in what was created in heaven and on earth. All the residents of the world are governed by one and the same star [i.e., destiny]. (*Tanna deBei Eliyahu Rabba* 2)

5 | ANIMAL WELFARE

There is an old legend that Moses, when a shepherd, found a stray lamb and tenderly carried the tired creature in his arms back to the fold. Thereupon a voice from heaven cried 'Thou are worthy to be my people's pastor'. From that day to this, when Tevye the Milkman converses feelingly with his weary carthorse, Jews have shown exceptional concern for animals. A thousand years ago there was even a rabbinical debate as to whether animals went to heaven. And, of course, the Ten Commandments guaranteed the right of cattle to a weekly day of rest, the first animal right in history.

That concern often puts me in mind of the high regard for animals shown in religions emanating from the Indian sub-continent, but the more I study of Jewish teaching, not least for the purposes of this book, the more delighted I am to have been born a Jew. Reading every Day of Atonement (Yom Kippur) the story of Jonah, the only biblical work with a touch of humour, I am constantly intrigued by the final words, 'and also much cattle'. We never forget the animals. Not surprisingly, Dr Lewis Gompertz, one of the founders of the Royal Society for the Prevention of Cruelty to Animals in Britain, was of Jewish birth.

What follows emanates from the Conference of Liberal and Progressive Rabbis, representing one of many religious strands in the Jewish world. Liberal in Britain is close to Reform in the USA. Reform in Britain edges nearer to Conservative in the USA. Orthodox here has shades to the left and right of it. A thousand flowers bloom in the constant Jewish religious debate, all seeking authority from the same basic Source.

61

Judaism moved early to protect the rights of animals and to the extent that Christianity and Islam express concern for animals, they have usually borrowed from Jewish sources. Yet Jews have been reluctant to proclaim this fact or to demonstrate how modern animal welfare societies have been inspired (albeit unconsciously or indirectly) by the teachings of our faith. Jews have cause to feel both pride and great sadness when they consider the exacting moral standards of Jewish animal welfare legislation and homily, and compare those standards with the neglect and abuse of animals so characteristic of most societies and legal systems, past and present.

Genesis 1:26 states that man may dominate all other creatures (which would appear inevitable, given superior human intelligence) but the dominion intended is a caring and responsible paternalism rather than a callous exploitation. In the Talmud (*Sanhedrin* 59b) 'dominion' is interpreted as the privilege of using animals for labour. The eleventh-century commentator Rashi understands the verse to mean that human dominion is granted by God on condition that animals are not abused. If, argues Rashi, we become unworthy of the trust placed in us by the Creator, then we will sink to a level lower than that of any animal.

Genesis 1:29 declares that man was initially meant to be vegetarian. Only after the Flood (contends Genesis 9:3) was human consumption of animals permitted and this was later understood as a concession, both to human weakness and to the supposed scarcity of edible vegetation. Horror at the slaughter of other creatures for food has, in our time, prompted a widespread return to vegetarianism, a position deserving of our attention and respect.

Genesis 24 relates how, when Eliezer went to Haran to find a wife for Isaac, he looked for a girl who would show kindness both to humans and to animals. After Rebecca had drawn water for Eliezer and his men she immediately watered their camels. Later Jewish teaching insisted that it should be the other way round and so the Talmud states, 'A man may not sit down to his own meal before he has fed his animals' (*Berakhot* 40a).

In Proverbs 12:10 we can read 'A righteous man has regard for the life of his animal', that is, he shows consideration for its needs and feelings. Such consideration should not be restricted to one's own animals for Exodus 23:4 teaches that stray animals must be taken care of and returned to their owners at the earliest opportunity. This applies even to animals belonging to one's enemy and the very next verse in Exodus 23 urges the offering of assistance to the fallen work animal of an enemy. Deuteronomy 22:4 demands that any fallen animal be helped to its feet. Exodus 20:10 teaches that animals must rest on the Sabbath day and the need for such consideration is repeated in Exodus 23:12 and Deuteronomy 5:14.

In Leviticus 22:27 it is stated that a young domestic animal may not be separated from its mother till at least seven days old and in Leviticus 22:28 it is prohibited to kill an animal together with its young, mainly in order to prevent the one witnessing the death of the other. Commenting on these two verses the twelfth-century philosopher Maimonides wrote:

> The pain of animals under such circumstances is very great. There is no difference in this case between the pain of humans and the pain of other living beings, since the love and tenderness of the mother for her young is not produced by reasoning but by feeling and this faculty exists not only in humans but in most living things. (*Guide of the Perplexed* 3:48)

Deuteronomy 22:6 and 7 forbid the capture of a mother bird together with her young or her eggs. If the young or eggs are required the mother must be absent when they are taken. In practice young fledglings were of little use to anyone so that this prohibition tended to protect mother and young together. Eggs were more likely to be taken but here the bond with the mother was much weaker.

Deuteronomy 22:10 states 'you shall not plough with an ox and an ass together'. These animals differ greatly in their nature, size and strength and it is consequently cruel to the weaker animal to yoke them together. The prohibition extends to the yoking together of any animals of unequal type. Deuteronomy 25:4 reads, 'you shall not muzzle the ox when he treads out the corn'. This prohibition was extended to include all animals employed in labour. It is sheer cruelty to excite an animal's desire for food and then prevent the satisfaction of that desire.

In Jewish post-biblical literature much is written about the need to spare animals from pain or stress. The Talmud forbids gladiatorial shows and hunting (*Avodah Zarah* 18b) so that bull-fighting, dog-fighting, cock-fighting and fox or big game hunting are quite abhorrent to the observant Jew. So too is the trapping of animals for such luxury items as fur coats, the mowing down of elephant herds for ivory or the merciless hunting of whales for the production of pet foods. Those who regard themselves as morally sensitive Jews are bound to avoid the purchase of trapped animal skins and furs, ivory or whale meat products. None may purchase an animal till he or she has first purchased the food for that animal to eat, declares Talmud Yerushalmi (*Ketubot* 4:8). A high percentage of the exotic creatures imported for sale as pets in Britain die of starvation before their crates or cages are opened. The observant Jew should avoid the purchase of imported animals, birds or reptiles.

A rabbinic parable suggests that the occupants of the Ark were saved only because of their compassion towards the animals in their charge. Often, the parable claims, they would deny themselves sleep at night in order to feed their charges (*Midrash Tehillim* on Psalm 37:1).

Much has been written and spoken against the Jewish method of slaughter but this method (known in Hebrew as *shehitah*) is actually designed to minimize animal suffering. As stated earlier, the consumption of animal flesh was regarded by Judaism as a concession to human weakness. Even so, the species of animal, bird and fish which may be consumed are severely restricted. The shehitah method renders an animal unconscious in a matter of seconds and it is doubtful if pain can be registered in such a short time. If it is, it can only be momentary and is as nothing compared to the life-long suffering endured by so many farm animals in our day. 'Factory farming' is an abomination and as the Talmud (in a summary of previous teachings on the subject) demands that animals be spared pain at all costs (*Baba Metsia* 31a–32b), the products of intensive animal husbandry must be considered as unsuitable for Jewish consumption. The Jewish consumer should purchase free-range eggs rather than battery eggs and avoid buying chicken or veal which derives from 'intensive farming'. To deprive God's creatures of sunlight, fresh air and exercise is utterly sadistic and it is against intensive animal husbandry, rather than against particu-

lar methods of slaughter, that the efforts of animal welfare societies ought to be directed. Shehitah is at least as humane as any
other method of slaughter but if one has serious doubts about the
morality of depriving other creatures of life then the honourable
course to pursue is that of vegetarianism.

Criticism may be levelled by Judaism against research laboratories where millions of animals are yearly tortured supposedly to
advance the frontiers of science. In many cases laboratory animals
are well treated and there is no question of cruelty being practised,
but in other instances it is difficult to avoid the suspicion that
laboratory personnel regard themselves as licensed to indulge in
sadism. At the very least one should distinguish between experiments intended to assist medical development and those con-
ducted for the benefit of commerce.

Tzaar baalei hayyim (animal suffering) is the rabbinic term
employed to embrace all Jewish law and lore concerned with
animal welfare. Such concern is global and includes reference to
animals in the wild, on farms, in laboratories, in zoos and circuses,
in pet shops and in private homes. The earth has been given by
God for the benefit of *all* creatures and we humans, as God's
stewards, must exercise restraint and recognize the rights of non-
humans, be they furred, feathered or scaled.

... and in recognition
of the work done
in this community by
our rabbi, ten trees
have been uprooted in
his name in Israel!

6 TU BI SHEVAT

A Happy New Year to all trees!

Philip L. Pick

Tu Bi Shevat (or Tu B'Sh'vat, the spelling doesn't matter) is the Jewish New Year for Trees. What a lovely idea, I always thought, when, on that day, as a child, I received a special gift of nuts, usually almonds, and we went out somewhere to plant a sapling. This happy celebration, un-biblical in origin, has become universal in the Jewish world, especially encouraged by a body founded in 1902 known as the Jewish National Fund, who have planted millions and millions of trees up and down the length and breadth of Israel.

Philip Pick, a pillar of Jewish vegetarianism in Britain, and Vice-President of the International Jewish Vegetarian and Ecological Society, has edited a delightful book called, appropriately, Tree of Life. *In it he enumerates those great Jews who have been vegetarians, from Israeli Chief Rabbi Kook to Shmuel Agnon, Nobel prize-winner.*

As a living testimony for his creed, Philip Pick feels strongly about issues. He hates hypocrisy and especially cruelty, but loves nature in all its forms. This comes through clearly in his short but illustrative essay.

Why is the festival of New Year for the Trees celebrated on the fifteenth of Shevat? It is because most of the annual rain in Israel falls before that date and thereafter the sap begins to fill the trees and their lives are renewed for another year of blossom and fruit. The *shekadiah*—the almond tree—is the herald of spring and it has pride of place in the celebration, for its rosy white buds are the first to blossom even before its leaves have sprouted. On this day the 'tithe' was reckoned, and Jewish farmers were obliged to take a tenth of their new fruit and crop produce to the Temple in Jerusalem.

In ancient times it was a custom to plant a cedar sapling on the birth of a boy, and a cypress sapling on the birth of a girl. The cedar symbolized the strength and stature of man while the cypress signified the fragrance and gentleness of woman. When the children were old enough, it was their task to care for the trees which had been planted in their honour. Today the main celebrations on Tu Bi Shevat is the tree planting ceremony, when pupils from every school assemble and follow their teachers into the countryside to plant young saplings. It makes the children aware of the need for re-afforestation and soil conservation to beautify the country.

In the Bible there are many lovely references to trees:

> For you shall go out with joy, and be led forth in peace; The mountains and the hills before you shall break forth into singing. And all the trees of the field shall clap their hands. Instead of the thorn shall come up the cypress. Instead of the briar shall come up the myrtle, (Isaiah 55)

> They shall sit every man under his vine and under his fig-tree. (Micah 4)

> Let the field exult, and everything in it. Then shall all the trees of the wood sing for joy. (Psalm 96)

> For the Lord thy God brings you into a good land, a land of brooks, of water, of fountains and springs flourishing forth in valleys and hills, a land of wheat and barley, and vines and fig-trees and pomegranates; a land of olive trees and honey. (Deuteronomy 8)

There is also a great deal in the Talmud on the care and love that should be shown towards trees because at the beginning of creation it is written 'And God planted the Garden of Eden'. Here are a few extracts.

> You shall not say: 'We shall dwell and not concern ourselves with planting', but as others planted for you, so shall you plant for your children.

> Once while Choni Hame'agel was walking along a road he saw a man planting a carob tree. Choni asked him: 'How many years will

it require for this tree to give forth fruit?' The man answered that it would require 70 years. Choni asked 'Are you so hale a man that you expect to live that length of time and eat of its fruit?' The man answered. 'I found a fruitful world because my forefathers planted for me. So will I do the same for my children.'

The tender roots of the fig split the hard rock of the crag.

The palm tree casts its shadow far from itself. The palm tree has no blemish. It provides dates for food, *lullavim* (branches) for Succot prayers, foliage for the Succah, fibres for ropes, leaves for winnowing purposes, and beams for supporting the ceilings and roof of the house.

It is forbidden to dwell in a city that has no garden in it.

This beautiful prayer is said at the tree-planting ceremony.

> Give dew for a blessing
> And cause beneficent rains to fall in their season
> To satiate the mountains of Israel and her valleys
> And to water thereon every plant and tree
> And these saplings
> Which we plant before thee this day.
> Make deep their roots and wide their crown
> That they may blossom forth in grace
> Amongst all the trees of Israel
> For good and for beauty.

The Sephardis have their own form of service which is printed in a small book called *Pri Etz Hadar* (Fruit of Goodly Trees) and a section of it comprises seventeen short chapters, each dealing with a different type of fruit grown in the Holy Land. At the conclusion of the reading of each of these chapters, the particular fruit described therein is eaten, and it is accompanied with four glasses of wine, one at the reading of each appropriate passage.

7 | SHEMITTA: A SABBATICAL FOR THE LAND

'The land shall rest,
and the people shall grow'

Shlomo Riskin

We turn from animals and plants to the land itself. Rabbi Riskin, writing in The Jerusalem Post, *Israel's leading English-language daily, refers to the weekly reading of a Torah portion in synagogue, which enjoins a rest for the land every seventh year. He draws some intriguing social conclusions from this law.*

Apart from not working the land, good for the soil, as best agricultural practice acknowledges, it would do the farmer no harm. He could go and study, enjoy a sabbatical himself. Extend the idea further from the land to urban industry, and it could help to reduce unemployment. People, stale in their jobs, could find a new sense of direction.

Of course university professors and other professionals still have their 'sabbaticals'. Intriguing how a law promulgated on Mount Sinai so long ago reaches forward in time.

Josephus, the Jewish historian of the first century, records that Alexander the Great dispensed with tribute from the Jews in the seventh year 'because they did not sow their fields'. It is thought Julius Caesar did likewise, although the Roman historian Tacitus saw the rest time as a sign of indolence.

When you come to the land which I give you, then shall the land keep a Sabbath to God. Six years you shall sow your field, and six years you shall prune your vineyard and gather in the fruit. But the seventh year shall be a Sabbath of solemn rest for the land, a Sabbath unto God. You shall neither sow your field, nor prune your vineyard ... (Leviticus 25:2)

For a long time, Jews living in the Diaspora were unable to return to the land of Israel, so the laws applying to the land itself were confined to holy texts and theories, without the possibility of practical application. But today whoever arrives in Israel is exposed to an entire array of agricultural laws unique to this land. The land of Israel, one soon discovers, has never lost its holiness. And even though there is no functioning priestly or Levitical class, certain aspects of *teruma* and *massar* still apply; thus a walk in the Jerusalem open-air market reveals store after store with rabbinic placards announcing that 'tithes' and 'offerings' have been taken from the fruits and vegetables on display.

But the most sweeping of all agricultural laws is the declaration that every seventh year the land must lie fallow, and whatever grows is deemed ownerless.

The last time we had a Shemitta year in Israel was in 1987.

In effect, during each seventh year in the Holy Land, it's forbidden to plant, plough or harvest, and the guiding principle is that if a plant or tree is in danger of dying—from lack of water or destructive weeds—we do what's necessary to maintain what exists. In effect, this is the beginning of Jewish ecology. But planting something new isn't allowed. The seventh year is a year during which the land belongs to no one, or everyone, and thus one can pluck fruit from a neighbour's orchard without worrying about trespassing or theft.

During the early years of the Zionist movement when young idealistic Jews arrived here to work and redeem the land, rabbinic authorities ruled that the life-and-death needs of society meant that this custom should be modified. Some of these measures are still allowed by the Chief Rabbinate of Israel, but many farmers and kibbutzim are paving the way towards total observance.

Shemitta, I believe, isn't simply an inconvenience and a twelve-month headache; but holds within it the dreams and visions of Judaism.

The first thing Shemitta teaches is that the land isn't ours, that physical objects do not really belong to us; we may think they do, and may even act as if they do, but they do not. Little by little, during a lifetime of seven-year cycles, the message sinks in. We are, in the ultimate sense, merely visitors on this world and on this land, and when the time comes to relinquish our physical possessions we must, whether we like it or not. Shemitta is an education in ultimates.

Second, Shemitta is to the world of space what the Sabbath is to the world of time; Sabbath is the seventh day, and Shemitta is the seventh year, and whatever the Sabbath teaches us about this world, ourselves, creation, freedom and God, so does Shemitta— not through the sanctification of time, but rather through the sanctification of space.

On the Sabbath, we acknowledge that our time doesn't belong to us, our employer or our government, but only to God. During Shemitta we acknowledge exactly the same thing. Everything is God's, and for one year—a complete cycle—we return the land to Him.

Third, calling this seventh year a Sabbath is a further reminder of the inherent freedom which Shemitta brings. On the Sabbath we are commanded not to work—even if we want to, even if the job excites us, even if there's a deadline. We are protected from growing enslaved to work, either from choice or desperation. And Shemitta means we must never grow enslaved to the land— whether we are tenant farmers, serfs or vassals. The seventh year, and not the pocketbook, is the limit.

Fourth, in biblical, agrarian Israel these seventh-year laws meant that the farmer, forbidden to work the land, could devote twelve months to study in the academies of Israel—something unique for the ancient world. Just as the land was allowed to absorb sun, rain and dew, the farmer for an entire year could absorb the sun, rain and dew of Torah study. Since the majority of people lived off the land, it's as if the entire population, and not just an elect few, had been awarded a study grant.

Of course, the farmer had to save enough during the six previous years to be able to devote the seventh to his own spiritual, religious and personal pursuits, and I think it's quite appropriate that it's the farmer, and not the craftsman, upon whom the commandment falls.

Furthermore, we can stretch the concept of fallow land to the realities of the industrial age. Why shouldn't the computer programmer or the TV anchorwoman get a chance to do what farmers in ancient Israel once did? The challenge here in Israel is to reach the point where the seventh year will be adhered to without resorting to rabbinical sanctions. God wants the land to lie fallow, and that must be our desired goal. But a greater challenge is to take post office workers or chemical engineers and give them a year of study. The entire society would be enriched by a sabbatical, not just farmers, not just teachers, and not just rabbis.

8 ‖ THE SOURCES OF VEGETARIAN INSPIRATION

Philip L. Pick

The Jewish view of the environment is not based on self-preservation or utilitarianism or even on a passing enthusiasm. It is based squarely on morality, on cleansing and purifying the souls of people, so that the world can ascend to the heights portrayed in prophetic visions.

This is the essence of Philip Pick's explanation of how flesh-eating began, as part of the human descent into violence, and why, with the end of violence to animals and eating the limbs of animals, we may transform ourselves and society.

I am reminded by this essay of the words 'He who is cruel to an animal will be cruel to a human being'. The laws of Moses, incorporated in Judaism, expanded by rabbinical commentary, by debate, by question and answer, over the millennia, are designed to limit acts of cruelty to all living things.

Perhaps today it could be said that he who is cruel to the environment is cruel not merely to the land and to other species, but also to himself and to his own society.

The Jewish philosophy of vegetarianism is a way of life that reaches back into the mysterious morning time of our history.

The story of man's first existence in the Garden of Eden may be based upon elemental truths, or may be just an ancient legend, but we believe it is a profound declaration of man's real relationship with his Maker, dealing with the essential nature of his being; it contains the seed of an eternal philosophy which points the way of human moral development and limits human ambitions. It guides his spiritual progress along the circumference of a vast circle until he reaches his starting point, and once again comes back to his original position as a caretaker of a garden, and the guardian of all that lives in it.

The first command is contained in Genesis 1:29, 30:

> . . . And God said, Behold, I have given you every herb-bearing seed, which is upon the face of all the earth, and every tree, in which is the fruit of a tree yielding seed; to you it shall be for food. And to every beast of the earth and to every fowl of the air, and to everything that creepeth upon the earth, wherein there is life, I have given every green herb for food: and it was so.

'In the primitive ideal age, as also in the Messianic future (see Isaiah 2), the animals were not to prey on one another' (Hertz).

On the completion of each phase of creation it is written 'And God saw that it was good' and on the sixth day 'God saw everything that he had made, and behold it was very good'. In total it was proclaimed 'very good' which indicates that the universe was as the Creator willed it, in complete harmony.

'This harmony bears witness to the unity of God who planned this unity of nature' (Luzzatto).

Until the time of Noah it was a capital offence to kill an animal even as it was to kill a man. This is confirmed by the statement in Genesis, 'To man and all creatures wherein is a living soul'. Note that the word 'soul' is applicable in the same way to man as to animals. Bearing this in mind many have wondered at the story of Cain and Abel, and in this context it becomes understandable. Why was the beautiful lamb which Abel slaughtered acceptable to God as an offering? And if this was so why did Cain, whose offering was scant in substance and begrudging in spirit, kill Abel? The story has two morals. First, that in giving, one should be generous and open-hearted and not count the cost. This Cain did not do, but Abel gave of his best. Secondly, the cardinal sin of killing a creature warranted capital punishment by the immutable law of retribution, and Abel paid the penalty. Because of the murder, retribution also overtook Cain. The era of violence and consequent retribution had begun and has developed even unto the present day.

75

The law 'an eye for an eye and a tooth for a tooth' has been much criticized by people who have no understanding of its awesome truth. It does not mean the return of injury for injury, but that judgement shall be applied with mercy and shall fit the crime. Dictators have been known to execute people for political views; this is not 'an eye for any eye', it is the absence of justice. This particular law is immutable and absolute and operates whether we like it or not. The story of Cain and Abel lives on today, where man and beast alike kill without cause, and eternal retribution is impartially exacted.

The time of Noah was at the end of the era of perfection. In Genesis 6 it is written:

> ... And it came to pass, when men began to multiply on the face of the earth ... God said, 'My spirit shall not always strive with man' ... and he saw that the wickedness was great and all flesh had corrupted his way upon the earth ... And God said unto Noah, 'The end of all flesh is come before me; for the earth is filled with violence through them'.

'Violence is described as "Ruthless outrage of the rights of the weak by the strong"' (Talmud).

According to the rabbis, God repented of his action in the same way as a parent will forgive and protect a child who has committed violence or even murder, and he put the rainbow in the sky as a promise never again to destroy the earth.

'For the imagination of man's heart is evil from his youth' (Genesis 8:21). The new era that followed accepted this fact. In the laws, from the time of Noah as in the consequent Hebrew laws given on Mount Sinai, statutes were not to be made which the people would not accept, as this would merely cause contempt for the law generally. Compromise was therefore essential in the hope that by accepting a code for living man would eventually return to his original self. At this time therefore permission was granted to eat flesh:

> And the fear of you and the dread of you shall be upon every beast of the earth and upon every fowl of the air and upon all that moveth upon the earth, and upon all the fishes of the sea ... every moving thing that liveth shall be food for you; even as the green herb have I

given you all things. But flesh with the life thereof, which is the blood thereof, shall ye not eat. (Genesis 9:2–4)

The celebrated Rabbi Hacohen-Kook, the first Chief Rabbi of Israel, wrote a clear-sighted treatise entitled *The Prophecy of Vegetarianism and Peace*, and in it he dealt with the above paragraph as follows: 'It is inconceivable that the Creator who had planned a world of harmony and a perfect way for man to live, should, many thousands of years later, find that this plan was wrong'. He refers to the dominion over the creatures as not being 'the domination of a tyrant tormenting his people and his slaves only to satisfy his private needs and desires. God forbid that such an ugly law of slavery should be sealed eternally in the word of God who is good to all, and whose tender mercies are over all his works.'

It is not often noticed or thought that eating flesh brings consequences. These words follow the permission to eat flesh: '... and surely your blood of your lives will I require ... at the hand of every man's brother will I require the life of man' (Genesis 9:5). So here is the permissive doctrine and its penalties. It has been proven that these penalties are inescapable and are evident in the present-day world.

When the Hebrews were eventually established in Israel, the law of Moses, which contains 613 precepts, was duly initiated. Notwithstanding that a mixed multitude of 200,000 accompanied the 400,000 Hebrews on their long trek from Egypt to the Promised Land, it was the most serious crime, after murder, to kill an animal outside the gates of the Temple, and carried the most severe penalty next to capital punishment. The great philosopher, doctor and Bible commentator of the twelfth century, Moses Maimonides, stated: 'The sacrifices were a concession to barbarism'. It must be remembered that child sacrifice was universal and as the story of the golden calf indicated, the people were surrounded by idol-worshipping tribes. The sacrifice of animals was to lead to the abolition of child sacrifice until eventually it led to its own abolition. Sacrifice is an essential part of the human make-up, as is evidenced today by the way people react in time of war and willingly sacrifice their lives. Primitive people could not understand any other form of worship, and today sacrifice is still required, but it is represented by charity and giving, which satisfy this instinct.

It was customary among all the tribes to drink the blood and cut the limbs from living creatures, with the false idea that they thereby took in the strength from the animal. This belief still holds good among some primitive peoples who continue this practice. The laws of Moses were designed to protect the animals from these cruelties, and to prevent the annihilation of the human species from the disease of flesh foods, by not consuming the blood 'which is the life thereof'. In this there was also a strong moral issue, and even today when an animal is slaughtered, some of its blood is buried in the ground and a prayer is said over it in order to remind the slaughterer that he has taken a life.

Orthodox Jews make a blessing for practically all benefits in life. There is a separate blessing for each type of food, but there is none for flesh foods—something that has been slaughtered cannot be blessed. There is a blessing on wearing new garments, but no blessing may be made over furs or other animal skins of any kind—you cannot destroy the works of creation and at the same time bless God for having them. There are blessings on seeing beautiful trees, famous people, thunder, lightning, etc., and the idea underlying it all is to acknowledge the supremacy of God and the dependency of man.

The festivals include Passover (Easter), Pentecost and Succot (Tabernacles). Some of these have been incorporated into Christian observance; the fast days, however, have not been adopted.

At Pentecost when the synagogues are decorated with fruits and flowers, no carcases of slaughtered creatures are to be seen. At Succot, when the little booths are erected, they are decorated with fruit and flowers; no bodies or portions of bodies are used as decorations. Even at Passover the paschal lamb is purely symbolic: there is no instruction to eat it other than on the first biblical Passover, and any food symbol can be used to carry out the ordinance that all generations shall remember the going out of Egypt; the departure from slavery to freedom.

On the solemn Day of Atonement, when all Jews fast and seek compassion from the Almighty for life and health in the coming year, no leather shoes should be worn in the synagogue. The reason for this is not humility but to avoid hypocrisy. It is not devout to pray for compassion when one has shown no compassion in daily life.

Great scribes, teachers and philosophers stride across the millennia of Jewish history, imbued with these teachings; many of them were vegetarian. Many followed the practice of sects in ancient Israel and helped to keep the flame of compassion from being extinguished. One of these sects, the Essenes, abjured all forms of flesh food and intoxicants. The Founder of Christianity may have been of this sect, and it is rather surprising that there is still discussion in vegetarian circles as to whether he was, in fact, vegetarian. The answer should be obvious, and stories such as the 'loaves and fishes' have other explanations.

It is interesting to note that a very much larger proportion of Jewish people are vegetarian than their neighbours. In many instances they take leading roles in furthering knowledge of this great subject. In Israel there have been three vegetarian Chief Rabbis in 25 years and over 4 per cent of the population are vegetarian.

The long winding road back to the Garden of Eden can now be clearly seen. May it be traversed ever more speedily and may the day not be far distant when the beautiful prophecy of Isaiah will be fulfilled:

> For behold I create new heavens and the new earth and the former shall not be remembered . . . and they shall plant the vineyards and eat the fruit of them . . . the wolf and lamb shall feed together and the lion shall eat straw like the bullock. They shall not hurt nor destroy in all my holy mountain.

9 THE ENVIRONMENT: ISRAEL'S REMARKABLE STORY

Aubrey Rose

What follows is a talk I gave to the Friends of Israel in Finchley, North London. It requires little introduction, except to say that, to the average Israeli, it must seem far too optimistic and glowing.

After all, Israel's citizens have to live with an acute water shortage, deterioration of conditions as massive demands are made by a flood of immigrants, whether from the republics of the former Soviet Union, Ethiopia, Albania, or elsewhere, industrial waste polluting streams, and the many problems the environmental flesh is perpetually heir to.

Yet living in that intense country often means an inability to see the wood for the trees. Great things have been achieved in the face of unbelievable odds. I really do believe that if peace one day descends on that area, the experience and expertise of Israel could be of enormous advantage to surrounding peoples, just as it has been to the 70 Third World countries it has assisted in agricultural and ecological enterprises. May the land, and all surrounding lands and their inhabitants, be blessed with peace.

It is noticeable that at various times certain issues surge to the fore in a country: human rights, race relations, industrial relations, inflation. There can be no doubt that in the last few years concern for the environment has become a dominant issue, not merely in Britain, but world-wide. This has been highlighted by an escalating number of international conferences and seminars in recent years: in Montreal, Norway, Malta, Holland, and in London. Today the words greenhouse effect, ozone layer, climate change, CFCs, sustainable development, rain-forest preservation, desertification, energy conservation, threatened species, green issues, green parties, have become part of our daily language, reflected in endless presentations on radio, TV, and in our press.

In the process I have observed two interesting points. First, where governments have complained about interference with internal affairs when confronted by Amnesty International strictures on breaches of human rights, today neither governments nor people have the slightest compunction about interfering in the internal affairs of other states on environmental issues. We have all bemoaned the cutting down of Brazilian and African rain forests and the effect on Yanomami and other indigenous peoples. Canadians have censured the USA for the spread of acid rain, while Scandinavians have done the same in regard to British and German poisonous industrial outpourings which have left trees dead and lakes rendered lifeless or dangerous. The lessons of Chernobyl and its vast international effects remain. National borders become irrelevant when environmental issues arise.

The second point is even more interesting. In 1988 UNEP (the United Nations Environmental Programme), held a regional meeting in Malta to discuss pollution in the Mediterranean Sea, dreadful in places, especially the Adriatic. I was in Malta at the time and sat in on part of the proceedings. Many nations were represented, including Israel, often nations who never sat around a table with each other, who had no diplomatic relations with each other. Again, concern for the environment pushed people and states together, irrespective of the state of diplomacy and political relations. The environment is no respecter of borders and governments. The environment, our natural world, existed before there were states and governments, and will continue to do so when there are no longer states and governments.

Yet the big issue today is what we, the organized, industrial and agricultural peoples, are doing, and how our actions affect our environment. What can we learn from each other, and teach each other? What examples are there of good and bad environmental behaviour? What are our reasons for caring for the environment?

There is much public criticism of Israel, some justified, some not. In 40 years the country has had five wars, and never any prolonged period of peace. Yet of what other country in the world can it be said that in the cause of peace it gave up an area of land larger than its own original size, namely the Sinai peninsula, indeed that it gave up valuable oil wells? It is a unique sacrifice in the search for peace. And against this turbulent background emerges a remarkable story of the Jewish people's relationship with the Land, *Ha'aretz*, a success story if ever there was one.

Throughout history whenever the hand of the Jewish people, as a people, has touched *Ha'aretz*, the Land has blossomed. Whenever the people have been separated from the Land, it has declined and drooped, become desolate and barren, as if it had lost heart. It is an unusual story. This love of the land has been more than just a search for political or physical asylum. There is a religious base, a historical base, spanning 4,000 years. The Chumash, the first five books of the Bible, has many injunctions about respecting the land. It should not be worked on the seventh day, it should rest every seventh year. Trees, even in the time of war, should not be cut down. Tree planting and well-digging go back to the patriarchs. The rabbis, for well over 1,000 years, from Hillel to Maimonides, from the Mishnah to the Talmud, developed these themes, with specific Jewish teachings on Green Belts, waste disposal, air pollution, alongside the doctrine of *bal tashchit*, 'thou shalt not destroy', almost an Eleventh Commandment. Thus when Theodor Herzl wrote his *Judenstaat* almost a century ago, and the Return to Zion, which is what Zionism is, gathered pace, there were already teachings and programmes in existence.

It is significant that one of the earliest acts of the Zionist Movement was to establish in 1902 the JNF, the Jewish National Fund, to renew and redeem the Land. It is significant too that its Hebrew name, Keren Kayemet, is a quotation from the Mishnah, written 1,700 years earlier. The JNF was an early Jewish 'World Wide Fund for Nature', responsible, as we saw in Chapter 2, for many tree-planting and land reclamation schemes. It is an extraordinary tale of Jewish devotion and sacrifice over these last 90 years, barely known outside the Jewish world.

These are some of the areas in which Israel has done so much: conservation of water, trickle drip system of watering plants, desalinization, sewage disposal, general irrigation procedures, including transportation of water; use of solar energy on a mass scale for heating and hot water, and on an industrial basis as a source of power; massive programmes of afforestation to stop soil erosion, build up top soil, tie down sand dunes, absorb pollution, screen off noise, beautify areas; the development of the kibbutz and moshav forms of agricultural settlement combined with light industry, a remarkable example which could particularly help Third World countries; reclamation of barren lands and removing hillside stones and boulders—a process in which Jews were happy to be the equivalent of hewers of wood and drawers of water; the restoration of sites of antiquity and of historical interest; the creation of wildlife and nature reserves, bird sanctuaries, scenic trails, the creation of biblical and botanical gardens and parks.

All this has been created in the face of wars and violence, the need for industrial development, factories and ports, massive housing demands—never more so than now—antagonism from neighbours, international commercial boycotts, defence demands on lands and resources of outrageous proportions, money shortage, and lack of raw materials.

It is a remarkable achievement, an outstanding achievement, in terms of ecology, whispered about by the Israeli Information service with an incomprehensible degree of under-statement.

It is said that the borders of Israel are clearly defined when seen from a satellite: on one side it is green, on the other, brown. No one visiting Israel can fail to be struck by the dynamism that exists, the outpouring of energy in so many spheres, often self-directed into political and religious arenas, but with a main thrust towards solving pressing problems no other state has had to face.

Consider the role of trees. In ancient times when something important happened an altar was built or a tree planted. The tree was real. It was also symbolic. We read of the tree of life, the tree of knowledge of good and evil. As we have seen, our festivals, Pesach, Shavuot, Succot, are all agricultural in origin, we even have a festival of trees, Tu Bi Shevat. How far the land of Israel had been neglected during previous centuries is shown by the achievement since independence in 1948. Since then in just 40 years over 180 million trees have been planted, whilst 11 million trees are being planted in 9,000 acres around Jerusalem. Similar programmes are in hand around other towns and cities.

85

Many people do not know that part of the recent Intifada has seen the burning of a million trees. The JNF propose to plant three trees for every one destroyed. Special forests arise yearly, commemorating individuals or events, e.g. President Herzog Forest, Soviet Jewry Forest. There is also a forest commemorating Thomas Masaryk, the first President of Czechoslovakia. There are over 100,000 acres of natural forest, 5 per cent of the country, 40 recreation areas and parks, 700 picnic and recreation sites, and at the same time scientific and research studies continue, especially in the universities, while nurseries for saplings abound. Trees planted have to suit the soil and climate. They may be pine trees near Jerusalem, cypress in the coastal plains, eucalyptus and acacia in the saline soil of the Negev and the Arava. So much has been achieved, so much more remains to be achieved, since over 50 per cent of Israel's land is still desert.

The same applies to *water*, a vital reality, but, like trees, also referred to in allegorical terms—*mayim chayim*, living waters, with endless references in the Bible, beginning with the very first Psalm.

There is the famous water carrier from north to south, from Kinneret and Jordan to the Negev, the building of dams to divert floodwaters, the creation of reservoirs, irrigation, the draining of marsh land, the creation of marine nature reserves along the Mediterranean coast and near Eilat. It has been a significant programme, with even more ambitious schemes being discussed, particularly new canal construction.

Not every scheme is a success. The draining of the Huleh wetland was accounted a great success. Yet the wetland had acted as a natural filter and cleansing mechanism for the Sea of Galilee, slowing down water movement, allowing nutrients to be used by the wetland species. Once it was drained the nutrient-rich water flowed freely into the Sea of Galilee, causing algae blooms, silting and related problems. Thus even experts learned the value of wetlands late in the process of land reclamation, and we all thereby learned of the delicate balance of nature. Yet the Huleh Reserve remains a great centre for bird species, just as Israel itself is a major point of migration for millions of birds.

For a tiny country Israel has an incredible variety of scenery and climate, from snow-capped mountains to the salt of the Dead Sea. It incorporates temperate, tropical and desert climates. Thus the Botanical Gardens in Jerusalem have sections growing plants, trees and bushes from every continent. Similarly, flora and fauna are extremely varied. There are 3,000 species of plant life, of which 150 grow only in Israel, 430 species of birds, 70 breeds of mammals and 80 of reptiles. Rare and wild flowers are a particular feature, especially in spring.

It is greatly to the country's credit to have achieved so much in conservation, with nature reserves spread throughout the land. Mount Meron Reserve near Safed in Galilee covers 25,000 densely-wooded acres, famous for its peonies.

Thus the principle of conservation and preservation pervades nature policy in Israel, not merely of the land, but also of famous sites of antiquity.

Israel is said to be an archaeologist's delight. It has a fascinating record in restoring historical sites, from Hezekiah's tunnel to Herod's Masada, Solomon's Hazor, the synagogue at Capernaum of Jesus' time to the synagogues of the sixteenth-century mystics in Tiberias and Safed. No one can fail to be impressed by the faithful and careful restoration of the 67 synagogues in the Old City of Jerusalem destroyed by Jordanians during their occupation. There is also the imaginative urban conservation in Jaffa and the dramatic buildings close to the Israel Museum, enshrining the Dead Sea Scrolls.

But inevitably there are increasing urban problems, problems of air and sea pollution from petrochemical industries near Haifa, similar problems and visual ugliness near Hadera, some morbid residential developments, an excess of urban litter, difficulties in road transport, often caused by the mercurial attitude of Israeli drivers—killing more Israelis than in all the wars—a tendency to urban sprawl from Tel Aviv, yet despite all this, positive achievements far outweigh the negative aspects.

There are in addition programmes of preservation, not of land or seas or historic sites, but of the animal world. For example the Ya'el Foundation—*ya'el* means 'ibex'—conserves nature and wildlife and initiates studies, such as that of the leopard in the Judean desert. A most imaginative scheme is the Hai-Bar Biblical

Wild Life Reserve, referred to as Noah's Ark in the twentieth century. Here are being preserved some of the world's rarest animals, in a landscape which is also being restored. The theme here, as in so much of Israel, recalls the words in our Sabbath service, when we ask God to 'renew our days as of old'.

Hai-Bar aims to restore endangered animals to their natural environment. All manner of animals have entered this Noah's Ark: wild asses, ostriches, gazelles, ibexes, the rare white oryx, the addax, as well as predatory animals like the wolf, the fox and the hyaena. In Israel the fox is preserved rather than hunted. This unusual centre of wildlife preservation shows respect for animals and serves as an example to surrounding countries where so many forms of wildlife have been hunted to extinction.

In all this programme of conservation there has been an attempt to balance science with aesthetics, history with development, human needs with the demands of the land and other forms of life. It is a difficult balance to strike, as we have seen in various recently revealed disasters: the industrial near-holocaust of Eastern Europe, the horror story of the Aral Sea, the merciless attacks on rainforests in South America and Africa, the widespread destruction of lakes, rivers and seas by acid rain, toxic waste and industrial pollution. Hence the sudden sense of urgency among industrialized nations. Hence, also, the fear of undeveloped countries that they will be left even further behind in economic development, selling their primary products merely to pay the interest on loans, condemning their exploding populations to hunger as they overwork the land and destroy forests for immediate need and, often, immediate greed.

It is encouraging that Israel has recognized the need to maintain this delicate and difficult ecological balance. After many years of effort 1989 saw the establishment of a new Ministry of the Environment, to which the functions of many other Ministries have been transferred. Almost all environmental laws are now administered by the new Ministry, from the Maintenance of Cleanliness Law to the control of hazardous waste sites. The environment has become a high priority issue, represented in cabinet. The Israeli newspaper *Ha'aretz*, suitably named, wryly commented that the new Ministry was the only one established in the last twenty years that was not superfluous. And the new Ministry will have to deal with the fight against waste pollution and control the transport of hazardous waste, now prohibited on roads adjacent to the Sea of Galilee.

The big plus is that environmental impact will be considered in every planning and development activity, often with environmental assessments similar to those required under European Community Directives.

In 1964, Israel's Nature Reserves Authority was set up to conserve natural reserves and assets, protect all forms of wild life, and promote environmental issues. They now administer 385 nature reserves covering 450,000 acres—a considerable amount of protected land in a small country like Israel. The NRA, as well as studying the interrelation between plants, animals and all living things, also recognizes the importance of development, yet tries to mitigate its ill effects by, for example, discussing with farmers the use of pesticides and examining streams to prevent/monitor pollution. The work of the NRA is looked at in greater detail in Chapter 11.

This then concludes our brief glance at the environment in Israel. There are problems. There will be greater problems, as its population expands to five million or more, for it is the incredible rate of population growth this century that is at the root of almost every environmental problem, an issue that every religious tradition will have to face squarely in the near future. Yet there are solid, positive achievements in Israel. There are comprehensive laws, planning expertise and public concern, and an increasing campaign to teach forms of social behaviour that respect the land, the towns and nature.

A nation, like an individual, is made up of different elements. A nation has its physical geography, just as a person has a physical body. That geography cannot be treated in isolation from the combined experience, knowledge and data accumulated in the state, in the same way as a body is affected by an individual's mind and intellect. These physical and mental elements cannot be divorced from each other. They continually affect each other. They include moral and social values as well as respect for learning and science, hope and a sense of purpose.

But a nation, like a person, has a third element. It has a soul, a spiritual content, that includes its recorded history and heritage, its language, and its unwritten memory, above all, its common faith. The Jew, be he or she secular or religious, is the inheritor of that spiritual impulse, fashioned in biblical times, developed by perceptive rabbis and teachers over the millennia. At the heart of that impulse is the Jewish belief in the unity of all things: man, animals, birds, nature, the total environment. The world is one, just as in the Shema prayer we state the Lord our God is one. And the Psalmist adds 'The earth is the Lord's and the fullness thereof'. We are his trustees, to look after and care for his creation.

If a people or a state can remain true to the best in its spiritual heritage that will become evident in its day-to-day relationship with its physical environment.

On this assessment Israel has done well. Her story should be widely told, and not remain hidden under a bushel. If the spirit of peace, mutual understanding and respect for life can begin to assert itself, amongst Israel's neighbours, and expand within Israel herself, then what Israel has achieved in her own environment can one day become an example and a blessing to peoples close to her and to peoples far removed.

10

THE REGIONAL AND GLOBAL SIGNIFICANCE OF ENVIRONMENTAL PROTECTION, NATURE CONSERVATION AND ECOLOGICAL RESEARCH IN ISRAEL

Uriel N. Safriel

With Professor Safriel, formerly Chief Scientist with the Israel Nature Reserves Authority, we enter a new aspect of the subject. No rabbis are quoted, no biblical sources rehearsed, no moral issues discussed. We are concerned solely with science and facts, the material of the expert. Yet the facts themselves are fascinating. Did you know that ecologically there are two Israels, one facing north and west to the Mediterranean, the other south and east to the tropical and desert lands?

Thus Israel is in a climate transition zone, containing many arboreal species in danger of perishing elsewhere as a result of the warming up of the planet. The world may yet turn to Israel to replace vital genetic material.

At the same time, because of its position between two worlds, it has become an advanced 'listening post' for global warming and climate change. Israel therefore has a global responsibility and takes on global importance.

Who would have thought it, such a sliver of land on the edge of endless Asia? Some who would were the mediaeval cartographers who placed Jerusalem at the centre of the world. According to Professor Safriel they may not have been far wrong. Read on, and be surprised.

Put me through to the Director of Research!

THE TWO 'ECOLOGICAL ISRAELS', DESERT AND NON-DESERT

Israel is a small country, mostly arid and poor in natural resources. It is subjected to increasing population pressures and experiences industrial development and economic growth, all of which have negative environmental implications. It is therefore not unexpected that environmental awareness and efforts to conserve nature have recently distinguished themselves on the national agenda. Yet the protection of the environment and the conservation of nature in Israel, and the scientific research of means and ways of achieving these, are not just beneficial to the State of Israel and its inhabitants, but have a strong regional, and even an important global significance.

The regional and global significance of Israel, in the environmental sense, stems from Israel's geographical position and from its historical background. With respect to geography, on the local scale, there are two different 'climatic' and 'ecological' Israels: the northern half of the country, which enjoys a Mediterranean climate and is 'green', and the southern half which is a desert. In each of these two very distinct regions there are natural entities and environmental features which constitute invaluable assets of global significance, which overshadows their local value.

THE WILD PROGENITORS OF CULTIVATED PLANTS— AN ASSET OF NON-DESERT ISRAEL

Historically, the 'non-desert' part of Israel has been the cradle of cultivation and 'domestication' of staple food plants. Often man has destroyed the wild progenitors of these cultivated species, to prevent their interbreeding and to maintain the characters beneficial for man, so that they are not 'diluted' by the wild, often undesirable features of the progenitors. But with respect to many species of food plants, notably the wild progenitor of the bread wheat, ancestors of cultivated species survive even now in the wild, within the Mediterranean region of Israel. Why are these wild plants an asset of global significance?

The cost of the ever-growing intensification of artificial selection of food plants for better and larger yields, is the loss of genetic plasticity. Genetic plasticity enables plants to withstand the vagaries of the ever-changing environment, and to maintain viable populations even when conditions become hostile. It has become increasingly clear that only the wild progenitors can be now used to invigorate cultivated plant species and make them resistant to newly occurring spontaneous diseases and to unprecedented changes in modern agricultural environments.

Not less than 25 species of wild progenitors of food plants live in natural habitats in the non-desert. Mediterranean part of Israel. They constitute an unprecedented rich genetic resource, a bank of genes for resistance ready for use by farmers the world over, yet restricted in its distribution to an extremely small part of the globe. Traditionally, such genes are preserved in 'gene banks', which are merely stores of seeds of the wild progenitor species. However, in 'gene banks' it is the legacy of the past which is preserved, whereas the wild plants in their natural habitats are continually exposed to the changing environment and their genetic constitution is constantly moulded by the forces of natural selection. The key to their success and reliability as a dynamic genetic resource for the feeding of humanity is their conservation in the wild habitat, i.e., in the proper management and the appropriate conservation of the ecosystems in which they live. Thus, environmental management and nature conservation of the northern, non-desert, Mediterranean part of Israel is actually an act of preserving assets of global economic value, most of which occur only in Israel. Means of wise management of this region should be improved in order to achieve effective conservation, and scientific research should guide these efforts.

THE DESERT AS AN ENVIRONMENTAL ASSET

The southern, desert part of Israel has an entirely different significance, which is again not just local, but is regional rather than really global. Traditionally, and usually justifiably, deserts have been regarded as inhospitable for man. Young Israel set out to 'conquer' its deserts and make them habitable. This early approach can be now viewed as rather naive; the desert can become habitable by man only with immense economic investment and a high permanent cost of transforming small parts of it to 'non-desert'.

Thus, the desert is largely unsuitable for living and will not provide solutions for human population growth and for the economic and environmental problems of developing countries. However, its recreational value and touristic significance for developed countries with increasing problems of urbanization, mega-cities, and social issues associated with growing leisure time is already, and will be much more in the future, highly appreciated. In this respect the Israeli desert (mostly the Negev, but also the Judean desert) is of great significance to Western, and in the future also to Eastern, Europe. This is because it is the most 'civilized' desert close to Europe, and within easy reach of many of its populations.

In the Negev, the Israeli urban population (and most of the Israeli population is urban), but also European tourists and holiday makers, can find a warm and sunny climate throughout the winter months; they are able to find solitude within inspiring landscapes of rare natural beauty, relatively close to where they ordinarily live, but so different in appearance and atmosphere from their living and working environment. Finally, in the Israeli desert, they are able to enjoy the amenities and conveniences of conventional tourist resorts, yet easily experience genuine wilderness at the same time. All these are guaranteed only if the southern part of Israel functions as a natural desert.

To function as a desert, the natural ecosystems of this region should maintain their typical structure and proper dynamic processes. Although they look lifeless from a distance, a closer look reveals rich fauna and flora which, together with the elements and intricately interacting with them, constitute the structure and function of the desert ecosystem. Harm only a few of these very peculiar species or some of these delicate interactions, and desert soils are eroded, air-borne dust increases, shade-providing trees die out, desert antelopes disappear—altogether the desert landscape loses its attractiveness, and subsequently also its regional value.

An illuminating example is provided by the political border between the Negev desert of Israel and the Sinai desert of Egypt. It runs as a straight line on the map, yet is also clearly visible on a satellite photograph: the Sinai side of it is remarkably lighter in colour than the Negev side. The dark colour of the Negev is contributed by a thin layer of microscopic, unicellular algae that carpets the desert surface. The algae secrete a mesh of fine but sticky microscopic threads, that hold together the desert's soil particles, to form a soil crust. On the Egyptian side of the border, the desert is overstocked with goats and camels. Their trampling tears off the delicate algal mesh, the algae die out, and the desert's soil crust breaks and disappears.

What is the environmental significance of the desert soil crust? The crust solidifies desert sand dunes and slows down their movements. It regulates soil moisture by reducing evaporation on the one hand, and increasing the flow of run-off water on the other hand. In these ways the crust, i.e., the algae, determines the patterns of distribution of desert plants and the animals that depend on them for food and cover, prevents soil erosion and the generation of air-borne dust, and thus maintains the integrity of the desert landscape.

These landscapes are controlled by a delicate balance between crust formation and crust breakage through digging by creatures such as desert ants, or trampling by larger desert wildlife species. The balance can be easily shifted by man. As well as overstocking, air-borne industrial pollutants are deposited on the soil by desert rains and poison the algae; military manoeuvres by heavy vehicles break the crust; and changes in the species composition and population sizes of desert animals determine the amount of vegetation cover and its spatial distribution. Much scientific research is required to understand these phenomena, to elucidate the mechanisms involved and thereafter to prescribe the correct amounts of livestock grazing, oversee the reintroduction of locally extinct wildlife, and plan proper road construction and management of tourist pressures, so that this unique natural resource, the desert, can be sustainably utilized.

THE SIGNIFICANCE OF ISRAEL AS A CLIMATIC TRANSITION ZONE FOR GLOBAL WARMING ISSUES

On a local scale there are two 'Israels', the non-desert Israel and the desert Israel. But on the global scale Israel as a whole has an unprecedented significance, totally independent of the issues just discussed. This is related to the issue of global change, or global warming due to the greenhouse effect, and to the fact that Israel as a whole is a country located at a global climatic transition zone. The 'desert Israel' is at the northern edge of the huge Saharo-Arabian desert region. The 'non-desert Israel' is at the southern edge of the Mediterranean climatic region, and the adjacent temperate European regions north of it. Taken together, both 'Israels' constitute a well-demarcated and clear transition zone on a global scale. This is demonstrated, for example, by the dramatic variation in rainfall, running from 20 mm to 900 mm across the mere 400 km length of the country. Also, and maybe even more importantly, the between-years variation in rainfall is greater in Israel than in the regions south and north of it. Thus, climatic transition zones are characterized both by rapid climatic change across their length, and by large climatic instabilities, as compared to the relative stability of core climatic regions.

Most species of plants and animals reach their limit of geographical distribution within climatic transition zones. Thus, species widely distributed over the whole Mediterranean climate region reach their southern limit of distribution in Israel, and species widely distributed all over the Sahara reach their northern limit of distribution in Israel. Organisms that live in climatic transition zones, i.e., at the limit of their species' distribution, differ from their counterparts living within the much wider core areas of their species' distribution. Transition zone populations are better adapted to climatic uncertainties and changes, than their core area counterparts. They are both sensitive to and tolerant of the occurrence and effect of climatic changes. This is because where changes are frequent, as in climatic transition zones, the forces of natural selection have led them to adapt to and prepare for change and to cope with it when it comes.

97

Bearing in mind these attributes of populations at climatic transition zones, these populations turn out to be invaluable tools for monitoring global change and predicting its environmental effects. Furthermore, they constitute a precious resource for repair of the predicted environmental damage associated with global warming. Due to the sensitivity of their species, changes in the structure and functions of natural ecosystems of a transition zone such as Israel can serve as very sensitive indicators of global changes. Similarly, because of their sensitivity these ecosystems are ideal for scientific experiments aiming at exploring the possible effects of global change on human environments, and in this way enable a prediction of these effects. Finally, due to their tolerance and hardiness, as well as remarkable genetic plasticity, the populations of plants and animals in such ecosystems are likely to survive global climatic changes far better than their counterpart populations in the much wider core areas of their distribution.

Global changes are already affecting our environment, and their future effect will soon be expressed in the gradual elimination of many plant and animals species from various regions of the globe. Policies of energy use leading to reduced emissions of greenhouse gases are already under way, and hopefully the trend of global warming will be slowed down, and eventually even reversed. However, by that time considerable environmental damage will have occurred. Yet it is highly likely that the core distribution areas will be more severely affected than transition areas. The role of transition climatic zones will then be in providing biological and genetic material for rehabilitation and restoration of damaged ecosystems in the core areas. Since Israel is, from the ecological standpoint, a transitional climate zone, its natural ecosystems, both its desert and non-desert ones, constitute a genetic and ecological resource of global significance, to be used for the restoration and rehabilitation of ecosystems, mainly in Europe, the Near East and the Middle East, following the forecasted damage caused by global change.

GLOBAL RESPONSIBILITIES OF CONSERVATION IN ISRAEL

To conclude, Israel is provisioned with an array of environmental assets of global significance: a large number of wild progenitors of cultivated plants guaranteeing the future of global agriculture; an unspoiled desert providing recreational options for heavily populated Europe; a living laboratory for monitoring global change and for scientific experiments aimed at predicting its environmental effects; and a natural open-door repository of genetic resources of high potential for restoration and rehabilitation of damaged global ecosystems. Conservation-oriented environmental research and the proper conservation of nature in Israel are therefore one of the more important global responsibilities of the State of Israel.

11 | NATURE RESERVES IN ISRAEL

Aubrey Rose

Over 2,000 years ago Rabbi Hillel stated: 'If I am not for myself, who then is for me; and if I am for myself only, what then am I, and if not now, when?' A typically Jewish statement, full of questions.

He probably only had human beings in mind. What the Nature Reserves Authority has done in Israel is to extend that statement beyond people, to all other forms of life, animals, fish and birds. Some of the examples of the rescue of dying breeds of animals, many mentioned in the Bible, are quite inspiring.

The world is full of destructive and constructive forces, which Judaism describes as good and bad impulses and urges. It is heartening to see in an area of the world where destruction has unleashed such misery these very constructive examples of Jewish care.

We were driving along the frontier with Lebanon when we had a puncture. No surprise to me after the rock-strewn hillsides over which our jeep had clambered. We got out while the ranger fixed the tyre.

A beautiful small yellow flower, a kind of miniature orchid, swung in the breeze on the edge of a long thin stem. Suddenly I let out a yell of excitement. There, just below the flower, a spider spun her web. Nothing unusual except that the spider was all sunshine yellow, every part of it, precisely the same shade as the flower. This is the kind of excitement a visit to any one of Israel's 385 nature reserves is likely to produce.

Our puncture repaired, we began to move off but not before a further discussion. Apparently the local military commander, a naturalist, had named every command post after a flower. When he had wished to cut down a small clump of trees for security reasons the Nature Reserves rangers had argued and argued until he relented and the trees remained. I wondered whether the rangers were influenced by the biblical injunction to preserve trees even in time of war. Or was it concern for the eleventh commandment, taught by the rabbis of old, 'thou shalt not destroy'?

Our ranger, a Druze, proud of his extensive reserve extending northwards from the Carmel National Park, was proud too of his son, recently accepted by the vastly over-subscribed parachute corps. He took us to his village where we enjoyed a superb meal. The Nature Reserves Authority is involved in the care of over 18 per cent of Israel's land surface, conserving natural values, a great diversity of species, soil, marine life, birds, vegetation, animals, even archaeological remains and historic sites, in a planned, scientific way.

All 300 of its staff have not merely knowledge but also love, love of everything that lives, and a deep concern for our human heritage. As the Authority states:

> Much of the world looks to Israel, the land of the Bible, as a trustee preserving the landscapes, flora and fauna of Scripture. Nature conservation in Israel applies modern conservation to the preservation of the gazelles and ostrich of the Koran, the wild lilies and foxes of the Christian Testament, and the many species which populate Jewish Scripture from the epic of Noah to the perceptive wonders of Job.

Not far from Haifa, in the quiet beauty of the lofty hills, is a reserve with a difference. In this region of the world, where the destruction of animals and birds is commonplace, the Hai-Bar, or Biblical Wildlife Reserve, is carrying out a programme of utter respect for the animal kingdom. It is trying to restore to the area the original wildlife as described in the Scriptures. We saw wild sheep and wild goats who, stage by stage, would be restored to their natural habitats. Even vultures were being bred, except that the five birds had so far failed to reproduce.

Especially wonderful was the saving of the fallow deer, a

beautiful, graceful animal, rescued from extinction in a James Bond-like operation. The original two males and four females had grown to a herd of 40. They would gradually be returned back to nature, to live and breed as originally intended.

A similar operation is in place at the other end of the country, Hai-Bar Yotvata near the Red Sea port of Eilat. There the Authority is restoring the ancient landscape, redeeming species threatened with extinction, each animal echoing in the mind a biblical phrase. 'My beloved is like a gazelle' sang Solomon in his Song. 'The high mountains are for the wild goats' declaimed the Psalmist. Jeremiah described the 'wild ass, used to the wilderness'. 'Who has loosened the bounds of the swift ass?' asked Job. The Book of Lamentations refers to 'the ostrich in the wildnerness'.

A short drive from Eilat, and all these animals can be seen, as well as the rare white oryx, saved with British Fauna Preservation Society help. An interesting side-light on this animal is that, in profile, it appears to have but one horn, from which, it is claimed, the legend of the mythical unicorn arose.

Each reserve has its own character, from the impressive stalactites and stalagmites of the Soreq Cave, part of the Avshalom Nature Reserve in the Judean mountains to the Nahal Me'Arot caves in the Carmel, revealing in its stones and fossils age after age, illuminated by a striking audiovisual presentation of the life of prehistoric man, shown in the dark caves themselves. Having seen such caves and the wild honey cones attached to their roofs, I had visions of the prophet Elijah or John the Baptist suddenly emerging, staff in hand. The land of the Bible brings to life the story of the Bible in a unique way.

For us, on this special occasion, each nature reserve visit meant a warm welcome from a new ranger, smart in his pale green shirt, eyes alert, peering keenly at every movement in the trees, every flight in the air. The rangers are very special people. Huleh Nature Reserve, north of the Sea of Galilee, close to the river Jordan, was another of these sources of excitement. Originally a malaria-ridden swamp, the Huleh had been largely drained in an enthusiastic drive to create much-needed farming land.

But, as we now know, from northern Greece to Somerset, you cannot drain wetlands without paying an ecological price. That price had been paid here in the loss of wildlife of all kinds: insects, animals, birds, vegetation, all creating an interlocking balance in nature. Man is the intrusive, often destructive, force, but here at Huleh, man is atoning and restoring the old conditions. And the animals and the fish and especially the birds are returning. It was the Huleh experience that led to the founding of nature preservation societies and eventually the setting up of the Nature Reserves Authority.

As we inspected the new lakes our childlike shouts of joy rang out as we spotted bird after bird: the grey heron, the egret, plovers, terns, even falcons, and the flashing, blue-winged kingfisher. Here, water buffalo, 180 of them, all females apart from three males, lolled about in the shallows. Their grazing habits are crucial in maintaining an even balance of vegetable life. Here, where climates of the north and south merge, papyrus reeds flourish alongside yellow irises, a meeting point for tropical and northern plants.

The Huleh is part of the great Rift Valley extending from Turkey to Tanzania. It is also part of the bird migratory route from the northern continents to Africa. With the gradual restoration of the Huleh swamp or wetlands the birds are now returning, en masse. Huleh and Eilat, important reserves, are now major stopping places for the great migration. Ornithologists from the four corners of the world come here to see sights of colour and splendour as hosts of birds rest on their long journey.

No one in this land shoots murderously at the birds, as happens in so many other countries, lands who shall be nameless but who should hang their heads in shame. Of course, inevitably there has to be some hunting and culling of animals, for the sake of nature's balance, but not for sport, or out of sheer wanton destructiveness.

The Holy Land, a mere sliver on the edge of Asia, is profoundly conscious of the realities of nature. It may be tiny, a milk and honey land, but it is supremely rich in species, particularly in the genetic parents of many of our common fruits and crops such as the pear, wheat, lentil, flax, melon, fig and almond. Israel could well become a world genetic bank for many species.

On a previous visit to the Red Sea we had inspected the Coral Beach Nature Reserve, a spectacle if ever there was one. On this occasion however our journey took us northwards. Crossing the Jordan we made for two of its tributaries. The word Jordan, according to rabbis of old, derives from two Hebrew words meaning 'the descent from Dan', and indeed the powerful Dan river cascades down in a fierce torrent, amidst jungle-like vegetation, set in a reserve ideal for rock-walking, with every facility for hiker and picnicker, archaeologist and botanist. How cold, pure and crystal-clear are these rushing waters which run alongside a variety of soaring trees, including an Atlantic pistachio reckoned to be 2,000 years old, but not as old as the Israelite tribe of Dan who settled here.

Our ranger drove us to another Jordan source, the Hermon river, gushing forth at the Banias waterfall, named after the nearby remains of a temple to the Greek god Pan. As at the river Dan, trout were plentiful. I even heard of the proposed introduction of salmon. Our Arab ranger invited us back to his home in a nearby village, where the coffee and the welcome was of the best.

In a few days, I had merely touched on the exciting world of Israel's Nature Reserves, so little-known in the outside world. The Authority has high aims and a clear philosophy of its own:

> All living organisms and natural phenomena have a right to exist— nature is beauty and a fundamental component of our physical and cultural existence—nature is quality of life—nature is the main resource for scientific progress which improves the well-being of humanity—human progress depends upon a positive relationship with nature—we must practise the wisest use of nature's richness without diminishing its reserves—protection of genetic diversity of wild species found in Israel could be critical for the future of human agriculture and animal husbandry—we are obligated to preserve this life-support heritage for future generations.

The Nature Reserves Authority, its director, scientists and rangers, may not know it—they might even blush if they heard it (they are that kind of people)—but they are doing God's work and preserving God's world in the best way they know how to. Their forebears in the Land, the prophets of old, the Psalmist, and the great rabbis of the past would indeed be proud of them, their present achievements and their future hopes.

12 | JERUSALEM'S BOTANICAL GARDEN

Watching the dream come true

Jerusalem is a world in itself. There is so much to see, such overtones of history, the scene of so many dreams. Not far from Israel's Parliament, the Knesset, close to a university campus, rolls a series of terraces, formerly a hillside bestrewn with rocks. Professor Michael Avishai, smiling, enthusiastic, guided us from South Africa to Australia, Europe to the United States, Asia to South America. So much in just 60 acres. Because of Jerusalem's unique position, trees and plants from every continent flourish in special sections of the Garden.

As an experimental garden it is unique, as a resource for city-dwellers, a boon. We saw a flower garden that could have blossomed in England. We tasted the sweet grass that could replace sugar cane. We inspected the dream that was coming true, the rich dream of botanic diversity, the application of science and experiment to everything that grows, sponsored by well-wishers in Israel and abroad, exchanging students with world-famous botanical gardens, another side of ecological Israel, the epitome of reverence for creation, another field of Jewish study.

What follows is based on an article in Leaves: News from the Jerusalem and University Botanical Garden.

In a land long known for its miracles, the greening of the Jerusalem and University Botanical Garden at the Givat Ram campus of Hebrew University seems to be another miracle. The remarkable transformation of this barren, rocky hillside into a Botanical Garden is beginning to arouse world-wide interest. The Jerusalem and University Botanical Garden is coming into its own, fulfilling the

105

prophecies of the founders who dreamed of a garden where plants from the four corners of the globe could be once more collected for a new beginning in the soil of Israel to add their measure of beauty for the new Jerusalem and to serve as a living symbol of peace for the present and future generations of this ancient land.

GEOGRAPHICALLY UNIQUE

Jerusalem is geographically situated at the border of the two largest plant regions of the world—temperate and tropical—and naturally supports four major plant phytogeographical regions. Many years of experience have shown that plants from vastly different climatic conditions can be successfully grown here. As the work continues, the gardens are fulfilling their primary role as an educational, recreational and tourist attraction, as well as a research centre and teaching facility, presenting within the aesthetic framework of a landscaped garden, the rich variety of plant life gathered from the four corners of the world.

The hillside has been landscaped with paths and major garden areas and the 60-acre site has been carefully laid out to accommodate the flora of ten major geographical areas: North America, Europe and Siberia, the Mediterranean, Central Asia, China and Japan, South America, tropical Asia, tropical Africa, South Africa and Australia and New Zealand.

Typical landscapes are being recreated here, utilizing plant life native to the areas represented and incorporating the hills, valleys and brooks to suggest their native environment. Thousands of plant species have already been collected and planted, most of them for the first time in Israel, and the greening of these botanical gardens is changing the face of the landscape on the eastern slopes of the Hebrew University's Givat Ram campus.

MAKING THE DREAM COME TRUE

The development of the Garden has taken place in three stages. In the first stage, the general scientific concept was developed by the

Scientific Committee. The concept was based upon plant geography, in which the Garden is divided into ten geographical sections. Following this, the well-known Jerusalem landscaping architect Shlomo Aronson was commissioned to provide a detailed landscape layout and, together with the members of the Scientific Committee, the detailed planting plan was prepared. The second and third stages are part of the ongoing process of developing the garden.

Stage two involves excavation, levelling, and filling of the raw, rocky hillside on which the Garden is built, creating the final contours of the land. Then, the paths and roads are paved, and the terracing is built. Only in the third stage is any planting done in the beds that were created during the first two stages, which are supported by a system of irrigation which is installed during this stage.

The garden is now open to the public.

Father reckons this Ark caper will make
the early editions of the Old Testament for sure!

13 NOAH'S SANCTUARIES

Liat Collins

Not far from Sidmouth, Devon, in one of England's beauty spots, is the Donkey Sanctuary. Here live ill and neglected donkeys, surrounded by care and compassion, mute animals rescued from the heartlessness of man who treats them as nonsentient disposable creatures. Alongside the Sanctuary is a centre for handicapped children. Thus humans and animals come to this home together just as Noah and his family entered the Ark alongside animals and birds who, seven by seven, two by two, found a haven from the Flood.

It is encouraging to see the establishment across Israel of Noah's Sanctuaries to provide a similar safe haven for dogs, cats and stray animals. And even more encouraging that these acts of kindness—gemilhut chasodim in Jewish terms—are based firmly on religions teaching.

My heart lifted when I saw reference to an Architect's Ecological Code of Ethics. As a lawyer who has practised in the criminal courts for many years I have often felt that architects and town planners should have been in the dock for creating heartless environments that brutalize their inhabitants. It may yet come to that. So architects and planners, beware.

Noah's Sanctuaries, established in 1990, is an umbrella organization formed to advance halachic environmentalism in Israel. It bases its philosophy on what it calls 'the most ancient source for environmental preservation activity, the Bible, which points out that man has been given full responsibility for the integrity of his environment and the preservation and development of the quality of life'. The commandments of letting work animals eat, rest and labour in equal pairs are fairly well known. But the Bible, Talmud and other major sources are full of less-known but equally compelling ideas about our relationship with animals and the environment, says Rosalyn Matzner, who carries out research for the organization. For example:

• Midrash Rabba posits that God tested Moses' ability to lead the people according to the way he treated the herds he was responsible for.

• In the Talmud, Rabbi Yehuda Hanassi is remembered as having suffered for several years for not having shown compassion to animals.

• Rabbi Yohanan went as far as to say that, had we not been given the Tora, we would have learned modesty from the cat. Similarly, the Maharal suggests that lessons can be learned from a dog 'whose name comes from his characteristics—*kelev* [dog] derived from *kulo lev* [all heart]'.

In fact, Judaism holds that there is no such thing as a bad dog, only a bad owner. Thus, when Noah's Sanctuaries approached the Chief Rabbis for their positions on different animal issues, the response was to assign all moral responsibility to the humans, who are obliged to show kindness: 'Again we return to the question of the morality of individual men and the society as a whole. Where is Israel going concerning human values? And this is the basic question of Tora and faith.'

Rabbis today believe that family pets have much to teach us. Haifa's Chief Rabbi, She'ar-Yashuv Haochen (who is a vegetarian), recommends a cat or dog to teach children compassion and tolerance.

Noah's Sanctuaries hopes to find human solutions to all animal and ecological problems by working within the halachic framework. Sometimes the decisions are easy, such as Jewish law's wholehearted condemnation of the poisoning of strays.

Other questions are not so quickly solved: until it is found to be halachically acceptable to neuter cats and dogs to avoid the suffering of numerous unwanted animals, the only solution seems to be 'sell' the animal technically to a gentile (much like *hametz* during Pessah) for the operation.

Acting in accordance with the commandment 'thou shalt not destroy', the group hopes to establish animal shelters (hence, Noah's Sanctuaries) in every city and town in the country—including smaller towns. These sanctuaries would act as educational centres to teach the basic elements of environmental preservation and care for animals, and to demonstrate methods of preservation. Plans include the recycling of food—using different leftovers to feed different animals.

But kindness to animals is only part of the picture. Noah's Sanctuaries has issued an 'Architect's Ecological Code of Ethics' to improve the whole environment and quality of life. The code is a list of such promises as:

> I pledge to make every effort possible to ensure the quality of life for man, domestic and wild life, as well as plant life in all urban, suburban and rural projects with which I am involved.

You say I cook too much for each meal. So tell me, how can I be economical with the leftovers if I didn't?

SECTION C
ACTION

14 THE USA: A JEWISH ECOLOGY GROUP

There are more Jews in the USA than anywhere else, although Israel is catching up fast. Behind them comes the former Soviet Union, and, a long way behind, France, Argentina and the Jewish communities of the Commonwealth. What happens in the United States therefore is important. Thus what Ellen Bernstein is doing in developing a nationwide ecology group is praiseworthy, significant and typical.

Typical, because, although we are a people of community, as Ellen Bernstein rightly states, we are also a people of individuals, individuals like Dr Ludwig Zamenhof, creating Esperanto in seeking a unified language, Albert Einstein, searching for a unified force field theory, Abraham and Moses declaring that the greatest unity was the Almighty.

Other Jews sought communal betterment. Karl Marx, with a rabbi in his background, sought economic betterment, Sigmund Freud sought mental betterment, even Prime Minister Benjamin Disraeli legislated for improved public health, a better environment through law, a secular halacha. In medicine the same applies. Communal health was advanced by Jonas Salk's polio vaccine, by Ernest Chain, the co-discoverer of penicillin, as much as by Maimonides many centuries before, exercising his medical skills for the benefit of the population of Egypt.

Ellen Bernstein follows therefore in a long tradition of caring, community, health, and individual enterprise.

What follows is based on articles by Rahel Musleah in the Long Island Jewish World, *and Michael D. Schaffer in* The Philadelphia Inquirer.

Ellen Bernstein has taken her two passions—ecology and religion—and merged them into her work: 'For me', she says,

> ecology and religion—and Judaism in particular—teach the same thing. The underlying principles are interdependence and cycles. Ecology is totally about community, and Judaism is totally about community. Ecology is also about the past, about evolution. So is Judaism. Both teach us a sense of place, give us a sense of humility. They speak to the same chord inside of me.

She prefers the world 'ecology' to 'environment', pointing out that *ecos* means 'house', so ecology is the study of the household:

> It teaches you how to relate to your planet and your house, not something outside you. 'Environment' suggests something outside of us—our environs, surroundings, rather than including humans in it.
>
> A lot of people know about environmental issues, but they need an understanding of the grandeur of ecological processes. Tu Bi Shevat is a time we should do that.

Ellen Bernstein is building the organization Shomrei Adamah on the success of her efforts over the last few years to revitalize the celebration of Tu Bi Shevat, the minor Jewish holiday in honour of trees. Tu Bi Shevat was not very widely celebrated, according to her, but she saw the festival as an opportunity to honour nature. She created a *seder* or home ceremony for Tu Bi Shevat and put together a *haggadah*, a collection of readings and prayers, for the festival. It has so far been celebrated in fifteen US states, from Maine to California.

The Religious Action Centre, the legislative branch of the Reform movement, asked Ellen Bernstein to put together an Earth Day pack for rabbis, including a liturgical piece, a sermon, a tree-planting ceremony to take place between Tu Bi Shevat and Earth Day (a period that has been targeted as a time for environmental awareness), recycling information and source material. The Conservative movement will have the option of sending the material out to its members as well.

The Jewish Theological Seminary is working on a study of the link between God, humanity and nature.

On a smaller scale, Shomrei Adamah is encouraging synagogues 'to make your place of worship truly a sacred place' by banning Styrofoam, setting up recycling centres, converting to recycled paper goods and using organic lawn products.

'I'm a real believer in small steps', says Ellen Bernstein. 'Making personal commitments precedes making political commitments. What people do in their homes and synagogues is a reflection of that.'

BROWNIE PRAYERS ON
THE ENVIRONMENT

Environmental concern has not been restricted to adults. Of course, the more you understand the intricacies of science, the more you understand the vagaries of climate change. But it needs little learning to perceive that wholesale destruction of forests has a direct effect on people, whether the native peoples in South America or Bengalis in Bangladesh. When you can't swim in Toronto's Lake Ontario because of acid rain and chemical accumulation you know at once what pollution is all about.

And children, even seven- and eight-year-olds, are no exception. Only they see things clearly and directly without the sophisticated hesitancy of older generations.

For the next contribution we have to thank the unusual combination of Lord Baden-Powell, my wife, and the enthusiastic youngsters of the Fourth Temple Fortune Brownie Pack in Golders Green, London, an area not unknown to the Jewish community. My wife Sheila runs the pack attached to a local synagogue, and hence mainly composed of Jewish girls. They compose their own prayers, in which the environment figures prominently.

Occasionally there emerge short pointed exhortations such as 'Thank you God, for helping me. Now go and help someone else', or the even terser request 'Please God, keep in touch' but usually they are petitions full of sincere feeling about people and nature. Here are a few, composed by the young Girl Guides we call Brownies.

117

Dear Lord
Thank you for nature, trees, flowers, insects and grass.
Help us to vanish the worst of the past.
Help us to learn from the bad things we have done.
Thank you God, very much. Amen.

Dear God.
As I write this prayer people are cutting down trees. This is not fair.
Trees play an important part in our life. I, if no one else, wish that this
was stopped.

Pollution is a big black cloud, we must stop it somehow.
So let's not drive in the car,
only use it for going far.

Dear God
Thank you for the world, the trees, the sun, the moon, the stars. Help us
not to pollute the world. Amen.

Dear God
Thank you for Brownies and thank Guiders for taking us on pack holiday. Thank you for the world you have created for us and try and make
countries at war have peace. Make people care more for others and stop
people polluting the world. Thank you. Amen.

Dear God
This is an anti pollution prayer. I pray that the people who use aerosol
stop using the silly stuff! We'll hold our noses until it stops!!

Dear God
Thank you for the world
With its flowers and trees.
Please help us to protect our environment.
And for everything living. Amen.

Dear God
Please help us to understand how important nature is. Don't let people
cut down trees. Don't let us pollute the air. Amen.

15 | ACTION ON THE ENVIRONMENT: A PRACTICAL GUIDE

Vicky Joseph

We are nothing if not a practical people. We may have visions and fly off into the clouds but our feet are always on the earth. So we come now to what we call tachlis, meaning, at the end of the day, what do we have to do, in rock-bottom, practical terms?

Vicky Joseph, Social Issues Co-ordinator for the Reform Synagogues of Great Britain, has compiled a list of 'do's' and 'don'ts' for individuals, businesses, local authorities, synagogues and Jewish organizations. (Incidentally I am always intrigued at the expressions 'Jewish synagogues', 'Jewish rabbis'; even Shakespeare refers to a 'Hebrew Jew'—tautology's finest hour.)

Apart from her work with RSGB, which has included producing significant pamphlets entitled Do Not Destroy My World, *Vicky Joseph has been a valuable member of the Board of Deputies Working Group on the Environment. She has set out in great detail in this article practical steps individuals can, and should take. Her contribution should be read and re-read, every reader asking the while 'how much have I done, and what should I do now?'*

In assembling our experts we were joined by one Jew who had had little contact with the community until sought out to advise on energy saving in buildings. He showed how, with the right design, materials, and equipment, enormous financial savings could be made in building and maintenance costs. What an opportunity for organizational treasurers of all institutions, in all religions, to turn their universal deficits into surplus, whilst being a friend of the environment (and their members) at the same time! Energy saving could ensure their re-election, should they so wish it.

The last few years have seen a phenomenal growth in environmental consciousness and a corresponding plethora of 'green' books, magazines, television programmes and consumer goods.

There is so much information, and misinformation, as manufacturers and retailers jump on the green bandwagon, that the principal difficulties in writing this section are to decide what is relevant, what information is true and which consumer goods are genuinely environmentally safe.

It is suggested that the bibliography at the end of this book is used as a reference for further reading. There are many areas in which action is necessary and the Government's White Paper *This Common Inheritance* has been used as a basis (now incorporated into the Environmental Protection Act 1990).

There are three general points to be made before examining specific actions.

1 In order to halt the inexorable environmental degradation that we are now experiencing, a fundamental change of attitude is required.

   This may be unpalatable but our greed and extravagant lifestyles are directly responsible for much of the damage to our environment. We need to re-examine our actions and priorities very carefully.
2 The environmental cost of one's purchases as well as the monetary cost should always be considered. Environmental responsibility need not cost the earth although lack of it may cost the Earth!

   In many instances financial savings are obvious, e.g. energy conservation, less use of the car etc., but even where environmentally benign goods are more expensive, e.g. organically grown food, high quality recycled paper, this is a temporary situation and as productivity increases to meet the growing demand, prices will come down.
3 Individual and community action are both of supreme importance. Community action can be empowering and can enhance the cohesiveness of the community, bringing people back to Judaism and to synagogue life.

   Individual action, though sometimes seemingly of little

effect, is equally valuable as *Pirke Avot* ('The Ethics of the Fathers') states: 'It is not for us to complete the work but neither may we desist from it'.

There are many biblical, talmudic and midrashic sources to support the idea of environmental protection and there are many serious problems to be tackled. What is urgently required is positive action by individuals, synagogues and other community organizations.

The Government's White Paper, in its chapter 'Action for All', sets out ideas for action under the headings 'Actions for Individuals', 'Action for Businesses' and 'Action for Local Authorities'. I have reproduced these ideas with some minor alterations and additions and have added a section titled 'Action for Synagogues and Jewish Institutions' which, although repetitive in some instances, I feel will be useful.

1 ACTION FOR INDIVIDUALS

As *householders* we can reduce energy consumption in a number of ways, e.g.:

— doing more to insulate and draughtproof homes;
— choosing the most energy efficient model when replacing boilers, fridges, freezers, washing machines, dishwashers, etc.
— switching off lights and electrical appliances, using low energy light bulbs and fitting timeswitches and thermostats on immersion heaters;
— setting temperature controls at reasonable levels.

All these actions not only save us money, but also reduce the emission of greenhouse gases which cause global warming. The Energy Efficiency Office (Department of Energy, 1 Palace Street, London SW1E 5HE) can provide more details. Householders can also take care to ensure that their homes and fires are properly ventilated to avoid pollution indoors. Using smokeless fuels reduces pollution outdoors.

As *travellers* our decisions influence greenhouse gas emission, as well as air pollution, and the quality of urban life. Ways to help include:

— adopting less aggressive driving habits to save fuel;
— keeping cars well-tuned;
— choosing cars fitted with a catalytic converter;
— buying more fuel-efficient cars;
— sharing car journeys with colleagues;
— using train or bus when we can;
— walking or cycling where it is safe;
— using unleaded petrol.

All this could save us money, as well as improving our environment and health.

As *shoppers* we can influence retailers and manufacturers, and take account of how the goods we choose will affect the environment as we use or dispose of them. We can help the environment by:

— buying recycled products whenever possible;
— not buying overpackaged goods, or goods in packaging that cannot be recycled;
— making our views known to retailers and manufacturers, especially those who do not provide sufficient environmental information to help us choose;
— not purchasing goods made of scarce materials—such as hardwoods, unless they come from a managed sustainable source.

As *consumers* we can reduce the need for waste tips or for new domestic waste incinerators, by reducing our waste. We can help by:

— recycling all our paper, rags, cans and glass, pressing our local authority or supermarket chain to provide better recycling facilities where necessary, e.g. for plastics;
— making the most of the produce we use—for example by reusing plastic carriers and returning egg boxes;
— disposing of fridges containing CFCs in accordance with local authority guidance;

— disposing of batteries, waste motor oil or household chemicals properly.

As *gardeners* we can:

— encourage urban wildlife with ponds, native trees, shrubs and flowers;
— use garden chemicals only when absolutely necessary;
— dispose of unwanted chemicals responsibly;
— recycle kitchen waste by making garden compost;
— choose alternatives to peat wherever possible, or use reduced peat composts for potting. The most promising of the alternatives to peat is made from coir (a by-product of coconut fibre);
— for mulching and soil conditioning use bark chips or leaf mould;
— buy bulbs from reputable dealers, making sure they've not been collected from the wild.

Planting trees can also help reduce global warming as they absorb carbon dioxide from the atmosphere.

As *good neighbours* we can help by:

— keeping noise to levels that do not disturb, controlling e.g. burglar alarms without a cut-off, lawn mowers etc.;
— putting litter in bins or taking it home, and keeping our dogs from fouling public places;
— improving the appearance of our home and garden;
— protecting any historic features of homes and streets;
— avoiding bonfires or smoking where it affects others.

As *investors* we can:

— seek information about the environmental practices of the companies we invest in, and make our views known.

As *responsible citizens* we can:

— take steps to inform ourselves of the facts relating to our environment;
— join organizations active in the environment;
— involve ourselves in local planning;

— alert the relevant bodies to possible breaches of planning or pollution controls;
— take care not to pollute the city, countryside, rivers or sea with non-biodegradable litter;
— encourage local authorities to provide facilities for recycling;
— make our views on the environment clear, not only by what we do, but also by telling local councillors, Members of Parliament, and Members of the European Parliament what we think.

As *parents* the Shema instructs us to 'teach our children diligently', this is arguably the most important action we can take. We can teach our children all these things, and be ready to learn from them, as they gain new environmental knowledge at school.

2 ACTION FOR BUSINESSES

Many businesses have realized that environmental issues have increasing influence on the competitive success of firms in the market, and are responding accordingly. So, apart from helping to achieve a better environment, good environmental management increasingly makes good business sense in its own right.

Business organizations such as the International Chamber of Commerce and the Confederation of British Industry have published guidelines on the integration of environmental considerations into business strategy, both generally and in specific areas such as waste minimization.

Many companies have developed a corporate environmental strategy to help ensure they achieve and sustain improved environmental management and maintain their competitive edge. Steps they can take include:

— reviewing all aspects of the business that may affect the environment, including products (design, development and production and the impact of their use and disposal), production processes (emissions, waste management and opportunities for recycling), energy use, building construction and management, transport and distribution, services and public relations (including the possibility of corporate sponsorship of environmental groups);

— identifying ways to improve the firm's environmental performance, including improvements in products, processes and support services (the options include minimizing the use of raw materials and guarding against accidents which may give rise to costs and pollution; for smaller firms expert advice may be available under the Department of Trade and Industry's Consultancy Initiative);

— assessing the costs of all the possible actions identified and the benefits to be gained;

— setting out in a policy statement the business's broad environmental objectives (such statements should say how the objectives are to be achieved, and how they affect organizational structure and responsibilities in the firm);

— involving staff at all levels in the environmental planning;

— reviewing progress and environmental audit to ensure that the environmental policies and objectives chosen are delivered effectively;

— consider publishing the firm's environmental performance in the annual report or elsewhere;

— keeping up to date with environmental issues relevant to the business, up dating and improving the action plan regularly.

3 ACTION FOR SYNAGOGUES AND JEWISH ORGANIZATIONS

As members of the above we can:

— inform ourselves individually and collectively about Jewish teachings on the environment;

— support local environmental groups by e.g. affiliating to their organizations or offering communal facilities for meetings;

— arrange for an energy audit and reduce energy consumption as suggested under the heading Action for Individuals;

— use only environmentally friendly building materials, cleaning materials, etc.

— avoid using non-biodegradable, non-recyclable disposable cups and plates for meetings and communal events;

— use recycled stationery, computer paper, toilet paper, etc.;

— avoid wasting paper. Use it sensibly, use both sides and use scrap for internal memos;

— recycle all waste paper from high grade computer paper to newspaper;

— recycle other items, for example wine bottles from kiddushim and aluminium drinks cans;

— make use of specific Jewish festivals to teach environmental responsibility e.g. when planting trees on Tu Bi Shevat. The three agricultural festivals Succot, Pesach and Shavuot, with their celebration of nature, are also opportunities to learn about environmental issues.

— remember that by complying with the religious requirements of Shabbat and festivals re walking to synagogue, refraining from using electrical appliances and so on you give the environment a day of rest too.

— consider the Jewish teaching *tsa'ar ba'alei chayim* (avoiding the suffering of living creatures) as related to the modern practice of kashrut. It could be argued that products from animals that have endured intensive factory farming methods infringe the spirit, if not the letter, of the Jewish dietary laws.

— set up an environmental monitoring group to ensure that agreed action is being maintained, new information is passed on, etc.

4 ACTION FOR THOSE INVOLVED IN LOCAL GOVERNMENT

Local authorities have always been in the front line in the protection and improvement of the environment, and in dealing with problems of pollution. Almost all local authority services affect the environment and environmental concerns need to be built into the way in which they are organized and administered.

Many local authorities have already reviewed their policies and programmes to ensure that all environmental concerns are properly taken into account in all aspects of their work. The local authority associations are developing advice for their members, based on actual examples, on ways in which authorities can develop their environmental role.

Issues which authorities should consider are:

— corporate management of environment issues within local authorities (how best to ensure that environmental concerns are properly taken into account by members and officers);

— reviewing and monitoring the state of the local environment (facts are the essential basis for action at local as at national level);

— reviewing policies and practices that affect the environment (authorities should examine how all their activities can influence the local environment);

— the preparation of environmental strategies or programmes, covering such areas as energy conservation, recycling, monitoring and minimizing pollution, environmental education, transport and planning policies, waste management, environmental protection and enhancement, environmental health, consumer protection, purchasing and own use of resources;

— involvement of the local community (local authorities are particularly well placed to work in partnership with voluntary groups, the public and the private sector on environmental issues).

16 | A SYNAGOGUE ENVIRONMENTAL GROUP

Sheila Chiat

The following contribution is a story of grass-roots efforts. How can a synagogue, or any institution for that matter, encourage members to understand and behave in a sensible way for the betterment of their environment? The writer tells of her efforts, of enthusiasm, of the problem of maintaining a sustained interest. She is fortunate in that she belongs to a synagogue group highly conscious of these problems, but at local level it remains uphill work. Persistence is needed.

There is the Jewish story of the criminal who pleaded with a judge that he could not serve the twenty-year sentence just imposed upon him. 'I know', replied the judge, 'but do your best.' More than that no one can do. The work is great and the labourers are few, but it is still not given to us to desist.

Also illustrated by this article is the function of a synagogue. The word itself is Greek, but in Hebrew it is a house of prayer, of study, of assembly, the latter referred to as Bet Knesset. In that capacity many groups form within the membership and an ecology group is appropriate for the many reasons set out in this book. To help the environment is to be Jewish in every possible way. How can we profess to be a holy people, if our thoughts are not pure and our actions not clean, and that is what ecology is all about. So carry on, Sheila Chiat and all who think and work like you in synagogues and institutions.

THE BEGINNING

The rabbi called an initial meeting, which planned to form a group, rather than a committee. They brainstormed about areas of concern and ways of working. They planned the introduction of the group and publicity over five weeks, linked to Tu Bi Shevat.

The aims of the group were:

— to ensure that the synagogue set standards, e.g. in waste disposal, recycled resources, energy saving;
— to increase the awareness of members with respect to the environment, through synagogue magazine articles, and publicity;

Introducing the group to the synagogue

This was prepared for by:

— a special service for Tu Bi Shevat;
— the rabbi's column and article in the synagogue newsletter;
— a display of children's work from holiday schemes and religion school;
— Oneg on a Friday night;
— an adult study lecture on 'Judaism and the Environment';
— a Pro-Israel Group evening on 'Israeli Botanical Gardens'.

The group was launched with a public meeting with a speaker from Friends of the Earth. The events were successful and about 50 people joined the group.

WORK IN THE AREAS OF CONCERN

Environmental responsibility in the synagogue

This was largely the concern of the steering group, and was not brought up at public meetings. The group tackled these areas: energy saving, materials to be used in the new building, water filters in the kitchen, the use of recycled paper, environmentally friendly cleaning materials, a synagogue no-smoking policy. The attitude of the synagogue administration was crucial for the success of these proposals.

The actual achievements included a voluntary no-smoking policy in official meetings; the administrator's assurance that cleaning materials and detergents were biodegradable; wine bottles were taken for recycling. But no can collection point was set up, and no notices about saving electricity were put up.

Increasing awareness among members

The synagogue newsletter has published articles regularly on areas of environmental concern. Other suggestions were made but not carried out: e.g. planting a tree in the synagogue grounds, selling environmentally friendly products on the premises.

Projects and open meetings

Links were made with local groups such as conservation volunteers, and some group members went on various conservation projects on Sunday.

An open meeting was held with the National Farmers' Union.

The Government and local authority were lobbied on nuclear waste disposal, water supply pollution, and public transport.

ONE AND A HALF YEARS ON

The steering group was largely a group of friends, who went rambling together on Sundays, and were active in other synagogue committees. They tended to do voluntary work, without trying to involve any other members of the synagogue. The number of open activities has been low, and the steering group felt that perhaps they should disband or change to a general Social Issues group. Questions remain about how best to organize to achieve the various purposes e.g., should there be a group to do political lobbying?

The group continues in existence, and is planning a display of publicity material and photographs in the synagogue foyer.

JUDAISM AND ECOLOGY

An overview of source material

Compiled by Sammy Jackman

I come now to the end of my work as your guide to the world of Judaism and ecology. Sometimes the words 'ecology' and 'environment' are used interchangeably, but there is, of course, a difference. Professor Nathaniel Lichfield, a distinguished member of our Working Group, describes environment as 'all those external influences which affect man and the human community, whereas ecology more strictly relates to the relationship of the environment to living organisms in Nature'. But be it environment or ecology, I hope the contributions have advanced your knowledge of the subject. They have advanced mine. Across thousands of years have come messages and teachings that remain as relevant today as when first uttered.

Time and circumstance have changed. Our world has changed. We may have changed, but the essential truth remains that we are all partners in God's creation. We have our rights but we have too our duties.

There is a Midrash or story of two men facing each other in a rowing boat. One is drilling a hole in the bottom of the boat beneath his seat. The other protests as the water pours in. The man continues drilling and replies 'But it is only under my seat'. 'Yes', replies his friend, 'but we are in the same boat, and if you don't stop we shall both be drowned.'

As I say farewell may I introduce you to a striking collection of sayings and quotations relating to the heart and soul of the Jewish approach, part of a golden treasury of wisdom and faith.

Thank you for continuing to this point. Thank God for preserving us in life until this time.

> I looked to the earth, behold! it was empty and void,
> to the heavens—their light was gone.
>
> I looked to the mountains, behold! they shuddered
> and all the hills broke apart.
>
> I looked, behold! no human being
> and all birds of the sky had fled.
>
> I looked, behold! the pasture was wasteland
> its cities torn down, because of the Lord
> because of his blazing anger.
>
> Jeremiah 4:23–26

ENVIRONMENTAL RESPONSIBILITY

How great are Your works, Lord; in wisdom You have made them all, the earth is full of Your possessions. (Psalm 104: 24)

All that (we) see ... the heaven, the earth, and all that fills it ... all these things are the external garments of God. (Shneour Zalman of Lyady, *Tanya*, chapter 42)

No mortal can in solid reality be ruler of anything ... God alone can rightly claim all things ... to this sovereignty of the Absolutely Existent, the oracle is a true witness in these words: 'And the land is not to be sold in perpetuity, for all land is Mine, because you are strangers and sojourners before Me' (Lev. 25:23). A clear proof surely that in possession all things are God's, and only as a loan do they belong to created beings. (Philo, Vol. 2, Loeb Classical Library, pp. 83, 119)

The earth and its fullness belong to the Lord, the world and those who dwell in it. Psalm 24:1

Rabbi Levi noted a problem: It says in Psalm 24, 'The earth and its fullness belong to the Lord', but it also says in Psalm 115:16. 'Heaven is the heaven of the Lord, but the earth he gave to the sons of man'. How do we reconcile these two statements? The answer is that before we have recited a blessing over something it belongs to the Lord, after we recite the blessing, and acknowledge its true ownership, then it is made available to us. Rabbi Chanina bar Papa says: Any benefit we derive from this world without saying a blessing, it is as if we stole it from the Holy One, blessed be he. *Berachot*

Why was the first human being called Adam? Rabbi Yehuda says: By virtue of the earth (Adamah) from which Adam was taken. *Midrash Hagadol Bereshit*

'And the Lord God took the man and put him into the Garden of Eden to work it and watch over it' (Genesis 2:15). The undefiled world was given over to man 'to work it', to apply to it his creative resources in order that it yield up to him its riches. But alongside the mandate to work and subdue it, he was appointed its watchman to guard over it, to keep it safe, to protect it even from his own rapaciousness and greed. Man is not only an *oved*, a worker and fabricator: he is also a *shomer*, a trustee who, according to halachah, is obligated to keep the world whole for its true Owner. Norman Lamm

When Rab Shlomoh Eger, a distinguished talmudist, became a Chasid, he was asked what he learned from Rab Menachem Mendel of Kotzk after his first visit. He answered that the first thing he learned in Kotzk was 'In the beginning God created'. But did a renowned scholar have to travel to a chassidic rabbi to learn the first verse of the Bible? He answered: 'I learned that God created only the beginning; everything else is up to human beings'. Quoted in article by Norman Lamm, *Nature*

In the hour when the Holy One created the first human being, God took the person before all the trees of the garden of Eden, and said to the person: 'See my works, how fine and excellent they are! Now all that I have created, for you have I created. Think upon this, and do not corrupt and desolate My world; for if you corrupt it, there is no one to set it right after you.' *Ecclesiastes Rabbah 7:28*

For thus says the Lord, the creator of heaven—he is God; who formed the earth and made it—he established it: he did not create it as a wasteland, but formed it as a place to live. Isaiah 45:18

It is clear that Judaism affirms without reservation that the world is God's creation and that whoever helps to preserve it is doing God's work. Louis Jacobs, *What Does Judaism Say About . . .?*

CONSERVATION—YOU SHALL NOT DESTROY

Those who chop beneficial trees will never be blessed in their work. *Pesahim* 50b

The lights of the world suffer because of the destroyers of beneficial trees. *Sukkah* 29a

It is forbidden to cut down fruit-bearing trees outside a (besieged) city, nor may a water channel be deflected from them so that they wither, as it is said: 'You must not destroy its trees' (Deuteronomy 20:19). Whoever cuts down a fruit-bearing tree is flogged. This penalty is imposed not only for cutting it down during a siege; whenever a fruit-yielding tree is cut down with destructive intent, flogging is incurred. It may be cut down, however, if its value for other purposes is greater (than that of the fruit it produces). The law forbids only wanton destruction. Maimonides, *Mishnah Torah, Book of Judges, Laws of Kings and Wars* 6:8–10.

(The) spoiler of all objects from which humanity may benefit violates this negative commandment (Bal tashchit ... You shall not destroy). *Shulhan Arukh Ha-rav*, para. 14

This prohibition of purposeless destruction of fruit trees around a besieged city is only to be taken as an example of general wastefulness. Under the concept of Bal tashchit (You shall not destroy), the purposeless destruction of anything at all is taken to be forbidden, so that our text becomes the most comprehensive warning to human beings not to misuse the position which God has given them as master of the world and its matter, by capricious, passionate or merely thoughtless wasteful destruction of anything on earth. Only for wise use has God laid the world at our feet when God said to Humanity 'subdue the world and have dominion over it' (Gen 1:28). Samson Raphael Hirsch

Because property is a sacred trust given by God, it must be used to fulfil God's purposes. 'Give him what is his, for what you are and what you have are his' (*Sayings of the Fathers*). Hence no man has absolute or exclusive control over his own possessions. One of the 613 commandments in the Torah is Bal tashchit, the prohibition against wastefulness.

A person who burns a garment or breaks a pot with the intent of destructiveness is considered as having violated the commandment because he has wasted that which did not belong to him alone. Similarly, the person who cuts down young trees in his own garden is subject to punishment because the fruit of the tree is not really his to destroy.

Waste is a loss not only for the owner but for his fellow men as well. Thus, a person is warned not to squander his fortune, even if he is well-intentioned and wants to devote his resources to charity, because his impoverishment would result in his becoming a burden to his fellow man. The underlying principle is that a man is held accountable to God for the responsible management of his possessions. Richard G. Hirsch

Rab Zutra said: One who covers an oil lamp or uncovers a lamp infringes the prohibition of wasteful destruction. *Shabbat* 67b

Garments may be rent for a dead person . . . But Rab Eleazar said: I heard that one who rends their garment too much for a dead person transgresses the command 'You shall not destroy'. It seems that this should be the more so in the case of injuring their own body. But garments might perhaps be different, as the loss is irretrievable, for Rab Jochanan used to call garments 'My Honourers', and Rab Hisda whenever he had to walk between thorns and thistles used to lift up his garments saying that whereas for the body, nature will provide a healing, for garments nature could bring no cure. *Bava Kama* 91b

One should be trained not to be destructive. When you bury a person, do not waste garments by burying them in the grave. It is better to give them to the poor than to cast them to worms and moths. Anyone who buries the dead in an expensive garment violates the negative mitzvah of Bal tashchit. Maimonides, *Mishnah Torah, Mourning* 14:24

THE LAND IS FOR OUR USE

And God said: Behold, I have given you every herb yielding seed, which is upon the earth, and every tree in which is the fruit of a tree yielding seed . . . to you shall it be for food. Genesis 1:29

God blessed them; and God said to them: 'Be fruitful and multiply, and replenish the earth, and subdue it; and have dominion over the fish of the sea, and over the fowl of the air, and over every living thing that moves on the earth'. Genesis 1:28

135

Rulership and power were given to you on land, to act according to your will with the beasts, crawling things, and all which crawls in the dust. And to build and to uproot that which is planted; to hew from her mountains bronze and that which derives from it. And this is what is included/permitted in the writing of 'all the earth'. Maimonides on Genesis 1:28

The phrase, therefore, refers ... to human conquest of the desert and constructive and civilizing endeavours to build and inhabit the world, harness the forces of nature for human good and exploit the mineral wealth around. Nehama Leibowitz on the above

To claim that (this verse) provides 'justification' for the exploitation of the environment, leading to the poisoning of the environment, the pollution of the atmosphere, the poisoning of the water, and the spoilation of natural resources is ... a complete distortion of the truth. On the contrary, the Hebrew Bible and Jewish interpreters *prohibit* such exploitation. Judaism goes much further and insists that humans have an obligation not only to conserve the world of nature but to enhance it because the human is the 'co-partner of God in the work of creation'. Robert Gordis on the above, from *Plaut*

POLLUTION

The quality of urban air compared to the air in the deserts and forests is like thick and turbulent water compared to pure and light water. And this is because in the cities with their tall buildings and narrow roads, the pollution that comes from their residents, their waste, their corpses, and offal from their cattle, and the stench of their adulterated food, make their entire air malodorous, turbulent, reeking and thick and the winds become accordingly so, although no one is aware of it.

And since there is no way out, because we grow up in cities and become used to them, we can at least choose a city with an open horizon ... And if you have no choice, and you cannot move out of the city, try at least to live in a suburb created to the northeast. Let the house be tall and the court wide enough to permit the northern wind and the sun to come through, because the sun thins out the pollution of the air, and makes it light and pure. Maimonides

The Talmud tells the story of a farmer who was clearing stones from his field and throwing them onto a public thoroughfare. A *chasid* (pious one) rebuked him saying, 'Worthless one! Why are you clearing stones from land which is not yours and depositing them on property which is yours?' The farmer scoffed at him for this strange reversal of the facts. In the course of time the farmer had to sell his field, and as he was walking on the public road, he fell on those same stones he had thoughtlessly deposited there. He then understood the truth of the *chasid*'s words: the damage he had wrought in the public domain was ultimately damage to his own property and well-being. Tosefta, *Bava Kama* 10:2, cf. *Bava Kama* 50b. (Paraphrased by Jonathan Helfand, *Judaism and Environment Ethics*)

Care was to be taken that bits of broken glass should not be scattered on public land where they cause injury. We are told that saintly persons would bury their broken glassware deep down in their own fields (*Bava Kama* 30a). Other rubbish could be deposited on public land, but only during the winter months when in any event the roads were a morass of mud due to the rains (*Bava Kama* 30a) . . . A tannery must not be set up in such a way that the prevailing winds can waft the unpleasant odour to the town. Louis Jacobs, *What Does Judaism Say About . . .?*, p. 130

A permanent threshing floor may not be made within 50 cubits from the city. No one may make a permanent threshing floor within one's own domain unless one's property extends 50 cubits in every direction, and it must be far enough away from the plantings and ploughed land of a neighbour for (the chaff) to cause no damage. Carcasses, groves and tanneries may not remain within a space of 50 cubits from the town. A tannery may be set up only on the east side of the town. Rabbi Akiva says '(The tannery) may be set up on any side except the west'. *Bava Batra* 2:8-9

The Mishnah says: If a person desires to open a shop in the courtyard, his neighbour may stop him because he (the neighbour) will be kept awake by the noise of people going and coming to and from the shop. *Bava Batra* 20b

REFORESTATION—RECLAIMING THE LAND

If you are in the midst of planting and word reaches you that the Messiah has arrived, do not interrupt your work; first finish your planting, and only then go out to welcome the Messiah. Rabbi Yochanan Ben Zakai

At the beginning of the creation of the world, the Holy One blessed be He began with planting first. For it is written: 'And the Lord God planted a garden eastward in Eden' (Gen 2:8). You too when you enter the land shall engage in nothing but planting. Therefore it is written: 'And when you shall come into the land, you shall have planted . . .' (Lev 19:23). *Leviticus Rabbah* 25:3

ENVIRONMENTAL LAND USAGE

Six years shall you sow your field, and six years shall you prune your vineyard, and gather in the produce thereof. But the seventh year shall be a Sabbath of solemn rest, a Sabbath unto the Lord, you shall neither sow your field, nor prune your vineyard. Leviticus 25:3–4

The Holy One blessed be He said to the children of Israel: 'Sow for six years and leave the land at rest for the seventh year, so that you may know the land is Mine!' *Sanhedrin* 39a

PRESERVATION OF SPECIES

The rabbis said: Even those creatures you deem redundant in this world like flies, bugs and gnats, nevertheless have their allotted task in the scheme of creation, as it says: 'And God saw everything that God had made, and behold, it was very good' (Genesis 1:31). Rabbi Aha ben Hanina explained thus: Even those creatures deemed superfluous in the world, like serpents and scorpions, still have their definite place in the scheme of creation. *Midrash Rabbah*

You must hate me, for you did not choose (to send a scout) from the species of which there are seven (that is, the clean birds of which Noah was commanded to take seven pairs), but from a species of which there are only two. If the power of the sun or the power of cold overwhelmed me, would not the world be lacking a species? *Sanhedrin* 108b (Noah's Dove)

No animal from the herd or from the flock shall be slaughtered on the same day with its young. Leviticus 22:28

Scripture will not permit a destructive act that will cause the extinction of a species, even though it has permitted the ritual slaughtering of that species. And he who kills the mother animal and its young on the same day, or kills the bird on the same day he takes the eggs, is considered as if he destroys the species. Nahmanides

It is forbidden to kill an animal with its young on the same day, so people should be restrained and prevented from killing the two together in such a manner that the young is slain in the sight of the mother; for the pain of the animals under such circumstances is very great. There is no difference in this case between the pain of man and the pain of other living beings, since the love and tenderness of the mother for her young ones is not produced by reasoning, but by imagination, and this faculty exists not only in man, but in most living beings. Maimonides

In the Talmud it is permitted to slay wild animals only when they invade the habitations of man; but to pursue after them in the woods, their own dwelling place, when they are not accustomed to come to human habitations, there is no commandment to permit that. Such pursuit simply means following the desires of one's heart. In the case of one who needs to do all this and whose livelihood is derived from it, of them we would not say that hunting is cruel, as we slaughter cattle and birds and fish for the need of man . . . But he who has no need to make a livelihood from it, his hunting has nothing to do with his livelihood, this is cruelty. Ezekiel Landau

THE PROPER USE OF TECHNOLOGY

In a remarkable passage we read that Turnus Rufus, a pagan Roman general, asked Rabbi Akiva which was more beautiful (or useful): the works of God or the works of humanity. Holding some stalks of grain in one hand, and loaves of bread in the other, Rabbi Akiva showed the astounded pagan that the products of technology are more suited for humanity than results of the natural process alone. So did Rabbi Akiva proceed to explain the commandment of circumcision; both world and humanity were created incomplete, God having left it to humanity to perfect its environment and its body. Similarly, the commandments, in general, were given in order that people thereby purify their character, that they attain spiritual perfection (M. *Tanhuma, Tazira*). Humanity, the created creator, must, in imitation of its Maker, apply its creative abilities to all life: its natural environment, its body, its soul. Rabbi Norman Lamm, 'Ecology and Jewish law and theology' in *Faith and Doubt*, KTAV, New York, p. 178.

WASTE CONTROL AND SANITATION

And you shall have a place outside the camp, where you shall go out. And you shall have a spade among your weapons, so that when you sit down outside you shall dig therewith, and turn back and cover over that which comes from you. Deuteronomy 23:13–14

It is forbidden to relieve oneself inside the camp or anywhere on a field. It is a positive commandment to prepare a special path for easing oneself there. For it is said: 'You shall have a place outside the camp'. Furthermore, it is a positive commandment for everyone to carry a spade as part of their war gear. One shall go out by that path and dig, ease oneself and cover up. As it is said: 'A spade shall be with your gear'. Regardless whether or not the Holy Ark travels with the troops, this must be the procedure to be followed. For it is said: 'Your camp shall be holy'. Maimonides, *Laws of Kings* 6:14–15.

BIBLIOGRAPHY

There is an immense range of books on Judaism although few on its view of ecology. The *Encyclopaedia Judaica* is a mine of information, as well as a source for further reference. Various Jewish religious organizations have excellent notes as to personalities and teachings in their Daily, Sabbath and Festival prayer books, as well as producing publications on social issues including the environment.

Inevitably I have to make a personal choice. I have suggested therefore a limited number of books which I have enjoyed and found instructive. Those, in turn, have their own lists of recommended reading, as well as numerous allusions to biblical, rabbinical and modern sources. The reader, having mastered these, will have encompassed a whole university of the Jewish faith, and will be informed of the rich diversity of Jewish historical and religious expression.

In addition I have added a few publications on purely ecological issues which will be of interest to those of the Jewish and every other faith.

Rabbi R. Brasch, *The Unknown Sanctuary: The Story of Judaism, its Teachings, Philosophy and Symbols*, Angus and Robertson, Sydney, 1969.

Rev Dr A. Cohen, *Everyman's Talmud*, J. M. Dent & Sons Ltd, London, 1932.

Lily Edelman (ed.), *Jewish Heritage Reader B'nai B'rith Book*, Taplinger Publishing Co. Inc., New York, 1965.

John Elkington and Julia Hailes, *The Green Consumer Guide*, Victor Gollancz Ltd, London, 1988.

Rabbi Dr I. Epstein, *The Jewish Way of Life*, Edward Goldston, London, 1946.

Rabbi Leonard B. Gewirtz, *The Authentic Jew and his Judaism*, Bloch Publishing Co., New York, 1961.

Martin Gilbert, *Jewish History Atlas*, Weidenfeld and Nicolson, London, 1969.

Rabbi Dr Louis Jacobs, *What Does Judaism Say About ...?*, Keter Publishing House Jerusalem Ltd, Jerusalem, 1973.

David A. Munro and Martin W. Holdgate (eds), *Caring for the Earth*, World Wide Fund for Nature, World Conservation Union and United Nations Environment Programme, Gland, Switzerland, 1991.

Chaim Pearl and Reuben Brookes, *A Guide to Jewish Knowledge*, Jewish Chronicle Publications, London, 1958.

Rabbi Jonathan A. Romain, *Faith and Practice: A Guide to Reform Judaism Today*, The Reform Synagogues of Great Britain, London, 1991.

Dr Cecil Roth, *A Short History of the Jewish People*, East and West Library, The Horovitz Publishing Co. Ltd, London, 1959.

Dr Cecil Roth (ed.), *The Standard Jewish Encyclopaedia*, W. H. Allen, London, 1966.

Rabbi Dr Jonathan Sacks, *Tradition in an Untraditional Age*, Vallentine Mitchell & Co. Ltd, London, 1990.

Rabbi Dr Norman Solomon, *Judaism and World Religion*, Macmillan, London/St Martin's Press, New York, 1991.

Colin Tudge, *Global Ecology*, The Natural History Museum, London, 1991.

Rabbi Eli Turkel, 'Judaism and the environment', *Journal of Halacha and Contemporary Society*, New York, no. xxii, Fall 1991.

White Paper, British Government, *This Common Inheritance: Britain's Environmental Strategy*, Her Majesty's Stationery Office, London, 1990.

Geoffrey Wigoder (ed.), *Jewish Values*, Keter Publishing House Jerusalem Ltd, Jerusalem, 1974.